Death, Culture & Leisure

EMERALD STUDIES IN DEATH AND CULTURE

Emerald Studies in Death and Culture provides an outlet for cross-disciplinary exploration of aspects of mortality. The series creates a new forum for the publication of interdisciplinary research that approaches death from a cultural perspective. Published texts will be at the forefront of new ideas, new subjects, new theoretical applications and new explorations of less conventional cultural engagements with death and the dead.

Published titles

Brian Parsons, *The Evolution of the British Funeral Industry in the 20th Century: From Undertaker to Funeral Director*

Ruth Penfold-Mounce, *Death, The Dead and Popular Culture*

Matthew Spokes, *Death, Memorialization and Deviant Spaces*

Racheal Harris, *Skin, Meaning, and Symbolism in Pet Memorials: Tattoos, Taxidermy, and Trinkets*

Dina Khapaeva, *Man-Eating Monsters: Anthropocentrism and Popular Culture*

Death, Culture & Leisure: Playing Dead

EDITED BY

MATT COWARD-GIBBS
University of York, UK

emerald PUBLISHING

United Kingdom – North America – Japan – India – Malaysia – China

Emerald Publishing Limited
Howard House, Wagon Lane, Bingley BD16 1WA, UK

First edition 2020

British Library Cataloguing in Publication Data
A catalogue record for this book is available from the British Library

ISBN: 978-1-83909-038-7 (Print)
ISBN: 978-1-83909-037-0 (Online)
ISBN: 978-1-83909-039-4 (Epub)

Printed and bound by CPI Group (UK) Ltd, Croydon, CR0 4YY

ISOQAR REGISTERED
ISOQAR certified Management System, awarded to Emerald for adherence to Environmental standard ISO 14001:2004.
Certificate Number 1985
ISO 14001

INVESTOR IN PEOPLE

Contents

Contributor Biographies

Vivian Asimos graduated with her PhD from Durham University and an MSc from Edinburgh University in the Anthropology of Religion. Her research interests rest at the intersection of religion and popular culture, with a special focus on storytelling and mythology. She is the Co-editor of the *Bloomsbury Reader in the Study of Myth* (2019) and is the founder of the *Religion and Popular Culture* Podcast.

Matt Coward-Gibbs is an Associate Lecturer in the Department of Sociology at the University of York and an occasional Lecturer in Religious Studies at York St John University. He is a Fellow of the Higher Education Academy and regularly teaches across undergraduate and postgraduate programmes in sociology, criminology and religious studies.

Ashley Darrow received his Master's in Gothic Studies from National University in 2017. Since, he has helped organise conferences on the intersection of Gothic studies and gaming and has presented papers internationally. He is the founder and co-host of the *Horror Vanguard* podcast.

Chloé Germaine Buckley is a Senior Lecturer in English at Manchester Metropolitan University. Her publications include *Twenty-First Century Children's Gothic* (2017) and numerous chapters and articles on all aspects of the Gothic. She also researches with the Manchester Game Studies Network.

Maggie Jackson began her career in Educational Psychology where her interest in working with bereaved children led to the publication of *The Teacher's Handbook of Death* (2002) with Jim Colwell. Following her doctorate exploring death in children's literature, she has begun to focus on death in picture books for young children.

Ewan Kirkland teaches Screen Studies at the University of Brighton focussing primarily on horror video games. He has published numerous papers and chapters on *Silent Hill, Resident Evil, Forbidden Siren* and *Haunting Ground* which have appeared in numerous journals including *Games & Culture, Convergence, Gothic Studies* and *Camera Obscura.*

Bethan Michael-Fox is a Doctoral student at the University of Winchester and an Associate Lecturer at The Open University. She has published on a range of topics that included representations of the undead in popular culture.

Christopher Partridge is Professor of Religious Studies in the Department of Politics, Philosophy and Religion at Lancaster University. He is the author of several books, including *High Culture: Drugs, Mysticism, and the Pursuit of Transcendence in the Modern World* (2018), *Mortality and Music: Popular Music and the Awareness of Death* (2015), and *The Lyre of Orpheus: Popular Music, the Sacred, and the Profane* (2013).

Benjamin Poore is a Senior Lecturer in Theatre at University of York. He has written widely on contemporary representations of the Victorians, literary adaptation and twentieth- and twenty-first-century playwriting. His books include *Heritage, Nostalgia and Modern British Theatre: Staging the Victorians* (2011) and *Theatre & Empire* (2016).

Paolo Ruffino is a Lecturer in Communication and Media at the University of Liverpool. His research spans video game culture, gamification, the quantified self and independent game development. He is the author of Future Gaming: Creative Interventions in Video Game Culture (2018) and editor of Rethinking Gamification (2014) and Independent Videogames: Cultures, Networks, Techniques and Politics (2020).

Jonathan D. Stubbs is a PhD student at Nazarene Theological College, University of Manchester. He is combining his lifelong love of video games, years of academic study of theology and fascination with all things 'apocalyptic' by researching the apocalyptic imagery in video games, focussing on Bioware's *Mass Effect* trilogy (2007–2012).

Andreas Theodorou is an independent scholar who began his studies on Nineteenth Century Gothic and the study of the mind, but later shifted focus to Gothic video games and the onus on interaction and immersion in contemporary narrative modes. He is the author and artist behind *Beyond the Darkness* (2018–2019), and a guest editor for the journal, *Alluvium*.

Solveiga Zibaite is a Social Anthropologist. Her PhD project, based out of the Centre for End of Life Studies at the University of Glasgow, is an ethnography of the Death Café movement in the United Kingdom. Her research interests include existential anthropology, online memorial cultures and sociological aesthetics.

Foreword

Since 2016, at the University of York, there has been a growing cluster of academics who are passionate, committed and focussed on matters of human mortality. Researching and talking about death, dying and the dead link us together and led to the formation of the Death & Culture Network (DaCNet), an interdisciplinary group of scholars and death industry professionals who wish to engage with death from a cultural perspective. As part of the DaCNet steering group, Matt Coward-Gibbs brings his wealth of experience to administrating the network ranging from his spates as an artistic director, youth theatre manager, research assistant and research centre manager. Add to this, Matt's teaching contribution at two universities and his stellar doctoral research, he is truly one of a kind and an emerging star into the research dominion of academia. Matt's vision and editorial work on this book only confirm this status and his value to the death studies community and beyond.

As part of DaCNet's various activities, its first annual symposium *Playing Dead* was held in May 2018. Memorably, the day had a bumpy start with technology problems leading to a last minute room change and a late start. Despite this inauspicious beginning and with participants being plied with coffee and biscuits, *Playing Dead* was a convivial event. Papers and discussion were insightful and thought provoking through an exploration of the intersections between death, culture and play. Play, in the context of this symposium, was used to broadly speak of leisure and recreational activities and, as such, not only encapsulated the playing of (non-)digital games, but also the consumption of cinema, literature and theatre. The symposium embraced the largely underappreciated phenomenon of play across disciplines by placing it within the context of death and culture.

This symposium was the catalyst for this book and draws together research passions that are united by play and focussed on death. As such it engages with the intricacies of playing digital games that inflict death in fantasy realms or are about memorialising and expressing grief. It examines scary folkloric-horror characters such as the Slender Man alongside death in the Pokémon universe. These death representations are held up in a complimentary, but contrasting, consideration of books that introduce children to death, the role of the dead and undead in television, the controversial relationship between music and suicide and even burial and exhumation in contemporary theatre. The significance of this work lies in the unity that Matt has managed to create through the interwoven thread of play – how we play, who we play and how we are playful in the context of death and the dead.

In this volume, Matt has succeeded in breaking new ground by drawing together interdisciplinary academics at different stages of their careers as well as integrating contributions from independent scholars. He has worked to encourage authors to convey their ideas across disciplines and to showcase their work and arguments to a high standard. The result is a collection of work that represents a range of voices that when brought together unite to provide a textured and vibrant contribution to death scholarship. Together these chapters bring strength and coherence to Matt's research interest and overarching research argument that play has intrinsic value in understanding the world around us in terms of community, labour, production and self-identity. Understanding and engaging with death through play is just one illustration of this and one that opens a wide realm to contemplate and investigate human mortality.

Ruth Penfold-Mounce,
University of York (UK)
@DeathandCulture
November 2019

Acknowledgments

They say it takes a village to raise a child; the same is true for an edited volume.

In the first instance, I would like to thank Ruth Penfold-Mounce for her ongoing support, encouragement, critical insights and unending thanatological enthusiasm. There are countless others who deserve recognition for their helpful advice and discussions along the way in the generation of this project. As such, I would also like to acknowledge David Beer, Rachael Burns, Alice Collett and Jack Denham as well as the wider publication team at Emerald for their care and assistance in the construction of this volume.

This volume would not have been possible without the tireless efforts of the twelve individuals who contributed to this collection. They have my sincerest thanks not only for their contribution, but for trusting this rookie editor with their work.

Finally, I would like to thank Emma and Ruby: there is no one else I'd rather play this game of life with.

Introduction: Death ≠ Failure

Matt Coward-Gibbs

> Death is not the opposite of life but an innate part of it.
> By living our lives, we nurture death
>
> Haruki Murakami (2003, 360)

In 2010, Julijonas Urbonas, artist and former amusement park employee, hypothesised a roller coaster that was designed to end its riders' lives. His *Euthanasia Coaster*, which begins with a two-minute ascent (for the purposes of personal reflection) and offers its passengers a chance to disembark before the drop, falls five hundred metres before entering into series seven 'teardrop shaped' concentric loops, maintaining the necessary G-force in order to euthanise its riders by starving both brain and body of oxygen (Urbonas, 2012). Following the demise of those aboard, the coaster makes a sharp turn, allowing for the disposal of the corpses, before returning to intake another set of passengers. Even in hypothetical discourse and playful engagement, separation from the corpse, and therefore the realities of death, is paramount.

Young and Light (2016, 61), in their recent article, note that spaces for the deceased in 'modern Western societies have been typified by a sequestration of death and the dead body'. This sentiment is not new, and still prevails throughout thanatological debate. Ariès, in *Western Attitudes Towards Death* remarked that when humanity entered the latter half of the nineteenth century 'death, so omnipresent in the past that it was familiar, would be effaced, would disappear. It would become shameful and forbidden' (Ariès, 1974, 85). This thesis of death's privatisation is interlinked with both the professionalisation and medicalisation of death through which a 'fundamental shift in the corporeal boundaries, symbolic and actual, associated with dead and the living' (Mellor & Shilling, 1993, 414) occurred. Physical spaces for the dead, such as cemeteries became a 'conscious attempt to relocate the corpse' (Rugg, 2000, 262) away from society emboldened by the growing fears of the corpse as a form of pollutant. These spaces of disposal became a 'boundary structure' (Rugg, 2000, 262) around the corpse. Changes, however, did not just occur with reference to the disposal but the location in which death took place. Medicalisation of the dying process

Death, Culture & Leisure: Playing Dead, 1–7

doi:10.1108/978-1-83909-037-020201007

moved it from the home to a clinical setting, both shrouding and making occultic the realities of death.

When death is obfuscated, autonomy of both the dying parties and witnesses is removed; therefore, societal understandings of death found primarily through either metaphor or adjacency. This means that, in most instances, we consume and understand death by proxy – at a position at least once removed from its realities. Jessica Fishman (2017, 51) provides a suitable example, in which she describes the way in which photojournalists are encouraged to 'hurry up and wait' when documenting death. For, in order to offer suitable coverage of death, images of mourners and memorials are infinitely more value than pictures of the dead themselves. Yet, Walter (2019, 389) goes as far as to contend that there is a 'new integration of the dead into everyday life', which he frames as the pervasive dead. However, this pervasive nature of the dead, Walter continues, can still exist with the thesis of death denial (Walter, 2019, 401; Becker, 1973; Ariès, 1974).

In the video game space, the notion of the *magic circle* is a somewhat contentious term (Huizinga, 1949; Salen and Zimmerman, 2003; Linser et al., 2008; Consalvo, 2009; Crawford, 2015, to name but a few).[1] However, in utilising the magic circle as a theme or tool, it becomes useful in allowing us to consider, more fully, the way in which death permeates leisure. In their article published in the *American Journal of Play,* Ken McAllister and Judd Ruggill (2018, 86) write that:

> [...] play and death are more than merely interchangeable in some ontological jiggery-pokery. They are, in fact, interrelated and interlocutory, dialogic in their ability to illuminate and inflect one [an]other.

It should therefore come as little surprise that there is a distinctive braiding between death, play and leisure.

Leisure itself can also be argued to lay in occultic realms. It is, after all, considered as the antithesis to work. In Western societies, in particular, individuals tend to be preoccupied on notions of 'production' and 'success' which are measured through various metrics, such as promotions at work or the amount of economic capital an individual has at their disposal (Bourdieu, 2004, 1977). Becker (1973, 11) contends, for example, that 'the fear of death must be present behind all normal social functioning'. He continues, suggesting that 'everything that man [*sic.*] does in his symbolic world is an attempt to deny or overcome his grotesque fate' (Becker, 1973, 27). If death sits at the root of human experience, even if that is but the management of terror (Solomon et al., 2015), then leisure is inevitably interwoven with death. Terror management theory, in essence, sits alongside Michaly Csikszentmihalyi's (2002) notion of optimal experience, or flow in which he contends that within each activity we undertake, we are part of a flow channel, carefully tracing a trajectory that is complex enough to avoid boredom, but not

[1]Both Jonathan Stubbs and I pick up on this issue in more detail in our respective chapters in this volume.

so challenging that it creates anxiety. Flow, Csikszentmihalyi argues, can be found in any state of life, including work, but prevails in both leisure and pleasurable situations.

However, leisure is deeply rooted with work. As Stebbins (2009, 3) remarks, across the academy, it is common for leisure to be viewed as 'little more than the purchase of a good or service'. In many instances throughout contemporary life, not just in recreation, we pay to consume. Life is expensive and so is death. There are many points at which the interface between death and capitalism is prevalent, especially in both formal processes of disposal as well as in popular media. The death of Maude Flanders in the popular US cartoon *The Simpsons*, for example, was primarily due to a pay dispute between voice talent Maggie Roswell and the Fox Broadcasting Company, when the latter refused to reimburse Roswell for her travel expenses (Blistein, 2013; Brownfield, 2000). Yet, in this instance, we also see a quasi-reanimation, with Roswell returning in order to voice Maude's ghost in later series following a settlement of the pay dispute.

This volume is primarily the product of the first *Death & Culture Network* annual symposium. The event that was held at the University of York in May 2018, brought together scholars from a variety of disciplines, backgrounds and career stages to consider the nexuses between death, culture and leisure. *Death, Culture and Leisure: Playing Dead* works within the realms of the once-removed from death. However, it does not accept that death should be approached through metaphor and adjacency as standard. It focusses on the way in which a variety of recreational, playful or pleasurable pursuits aim to grapple with the complex issues of death, dying and disposal. This volume aims to accept grey areas and work with contradictions in order to present a series of innovative approaches to contemporary thanatological debate. Divided into four sections this volume approaches: understandings of death in recreational landscapes; encounters with the gothic in playful environments; spaces for play with monsters and the realities of combining leisure and death.

Accompanying each chapter in this volume is a series of related materials for further consideration. These 'interactions and sensoria' offer the opportunity for further engagement whether that be through additional academic consideration, or further media texts to consider. Although this volume operates across four parts, it is somewhat difficult to separate them, meaning that the lines between different enquiries are often blurred. This is not meant to be a binary and, as such, readers are encouraged to dip in and out, skim or, if you please, read it entirely in reverse.

Playing with Understandings

The first section of this volume considers the way in which leisure and recreational activities allow us to play with understandings. To challenge ideas, critique and construct new rubrics. In the first chapter of this text – 'Nonhuman Games' – Paolo Ruffino questions and complicates the playing of games both based on and during the dawn of a new epoch, documenting and analysing the value of gaming media in the context of mass-extinction. As well as considering games which have engaged with post-Anthropocentric encounters, including *Horizon Zero Dawn* (Guerrilla Games, 2017), he moves to unpick those games which exist without the need for human intercession. In using this contemplation of 'idle' games, he maps a framework for non-human gaming, questioning the paradoxical notion of games that are both made, and yet for, no one. Stepping from Ruffino's consideration, Ashley Darrow's 'Peaceful in Death', shifts the tone somewhat in order to illustrate what is lost in conceptualising death in playful environments when our frame is locked firmly on meditative encounters which directly consider death. In his exploration of both explicit and implicit engagements with death within the infamous Japanese roleplaying game franchise, Pokémon. Darrow questions the way in which we consider Pokémon as a discourse around death, arguing that we should apply the same framing to it as we do to other cultural or media texts.

In 'Staying Dead', Benjamin Poore considers the way in which both burial and exhumation (by both the buried themselves and others) can be aligned with notions of nationhood and belonging. In his rich consideration of place and the way in which the playing-dead body is engaged with in live theatre, Poore invokes notions of folk horror in order to explore what a combination of theatre and nationhood can present in uncertain political times. The final chapter in this part – 'Death, Playfulness and Picture Books' presented by Maggie Jackson subsequently moves from performance to pictures in her discussion of the presentation of death in books for children. Jackson devotes time in this chapter to considering the directness of death's portrayal across a sample of children's picture books, drawing on the work of her PhD thesis. It is in their directness, Jackson argues, the children's picture book that deals with death does not have to sacrifice either artistic or literate merit, and can, at the same time, offer a profound route for both discourse and understanding.

Gaming Encounters in Gothic Environments

The second part of this volume presents a trio of encounters with death that are framed and centred through their Gothic nature. Jonathan D. Stubbs' 'Living and Dying in the City of the Damned' begins this section by considering mechanics of perma-death and Gothic intertextualities within Rogue Factor's (2015) tactical roleplaying game *Mordheim: City of the Damned*. By offering an intensive close reading based on the framework of Bizzocchi and Tenenbaum (2012; Bizzocchi, 2007), Stubbs considers the way in which a game's landscape can provide a fruitful location for the consideration of digital life lost. He considers the way in which perma-death in the context of *Mordheim* acts as a space for rumination on both insignificance and impermanence. The theme of space is also present within 'Prepare to Die', Andreas Theodorou's discussion of the *Dark Souls* trilogy (FromSoftware, 2011, 2014, 2016, and 2018 (remastered)). In this chapter, Theodorou contends that the *Dark Souls* trilogy has complicated the rubric of survival horror, calling for game texts to be read along range of experiential factors. Through the way in which the series demands that players themselves hunt for narrative congruency, Theodorou suggests that death, within the game, can be seen as a point of liberation and a realm to mark a players development.

As a way of rounding out a trio of gothic framed engagement 'He Died a Lot' from Ewan Kirkland offers an discussion of death in Giant Sparrow's (2017) acclaimed walking simulator *What Remains of Edith Finch*. In this consideration of uninhabited space, which notably chimes with the work of Ruffino and Stubbs, Kirkland figures death as a preoccupation for both the Gothic as well as video games, considering the morbidity of virtual space. By questioning the symbology of death within *What Remains of Edith Finch*, and the way in which the player is rendered as witness to each family member's demise, Kirkland draws out the way in which death is played with within this suitably Gothic virtual world.

Frolics with Monsters

The third section of this volume turns from the Gothic to consider the monstrous. In the first chapter of this part, Bethan Michael-Fox's chapter 'Dead Chatty' focusses on the implications of giving the undead voice. By constructing a typology, Michael-Fox unpicks the interfaces of undead dialogue in a range of contemporary media texts; ruminating on the pleasures and perils of undead articulation for both the living and the (un)dead. She considers the place of giving the undead voice in contemporary society, situating it in the wider cultural zeitgeist of the denial or possible non-denial of death, and the greater need to give voice to discourse about death, dying and disposal.

Vivian Asimos, in her chapter 'The Slender Man', asks if monstrous creation can be playful. Building on her work in contemporary myth and mythmaking, Asimos traces the way in which a monstrous legend began, quite innocently and playfully, within an online forum. In framing the generation of the Slender Man mythos alongside anthropological considerations of play, she figures the way in which society, the monstrous and the dead interact.

For the final chapter of this section, Chloé Germaine Buckley returns to considerations of the undead, in particular the zombie, in her chapter 'Gameful Interactions'. Through considering the ludic zombie, in relation primarily to augmented reality media sources and literature, Germain Buckley considers the way in which gaming's meritocratic norms co-exist with death. By considering the reader as a player, a complicated notion between goal-orientated play and death is discussed.

Performing Playful Realities

In this final section, the boundary is blurred somewhat between realms. Opening this section, Solveiga Zibaite's 'The Jovial Aesthetics of the Death-Positivity Movement' considers the role of humour within activism. By working with Bakhtinian notions of the carnivalesque, Zibaite encounters the way in which members of the Death Positivity Movement, particularly *The Order of Good Death*, known fondly as *deathlings* engage with humour as part of their social activism. In dissecting the movement's online practices, particularly focussing on its selling of apparel and the presence of its founder, Caitlin Dougherty, Zibaite considers the inherent 'style' of death positivity.

Following from this encounter, in 'Some Games You Just Can't Win' I present a case of a game based on the real-life experiences of a family coping with the death of a child. In a consideration of *That Dragon, Cancer* (Numinous Games, 2016), I focus on digital gaming spaces purpose-built for memorialisation, and the way in which crowdfunding has allowed *That Dragon, Cancer* to become a co-opted memorial.

The final chapter of this volume – 'Suicide, Angst, and Popular Music' – is presented by Christopher Partridge. Beginning with the understanding that mortality and music have always been interconnected, Partridge considers the way in which music offers not only a space for the musician to contemplate mortality, but also an opportunity for their listenership to do the same. In focussing on suicide, and the work of Durkheim, Partridge contemplates the relationship between suicide, mortality, self-harm and popular music.

References

Ariès, P. 1974. *Western Attitudes towards Death: From the Middle Ages to the Present*, Translated by P. Ranum, Baltimore, MD, The Johns Hopkins University Press.

Becker, E. 1973. *The Denial of Death*, New York, NY, The Free Press.

Bizzocchi, J. 2007. Games and narrative: an analytical framework, *Loading – The Journal of the Canadian Games Studies Association*, 1(1), 5–10.

Bizzocchi, J. and Tanenbaum, J. 2012. Mass effect 2: a case study in the design of game narrative, *Bulletin of Science, Technology & Society*, 32(5), 393–404. https://doi.org/10.1177/0270467612463796

Blistein, J. 2013. The Simpsons' to kill off major character, *Rolling Stone* [blog], 1 October. Available at: https://www.rollingstone.com/movies/movie-news/the-simpsons-to-kill-off-major-character-190587/

Bourdieu, P. 1977. *Outline of a Theory of Practice*, Translated by R. Nice, Cambridge Studies in Social and Cultural Anthropology 16, Cambridge, Cambridge University Press.

Bourdieu, P. 2004. The forms of capital, In *Handbook of Theory and Research for the Sociology of Education*, Ed. S. J. Ball, Translated by R. Nice, revised ed., pp. 15–29, New York, NY, Routledge Falmer.

Brownfield, P. 2000. Simpsons' actor disputes fox on reason for leaving. *Los Angeles Times*, 4 February. Available at: https://www.latimes.com/archives/la-xpm-2000-feb-04-ca-60883-story.html

Consalvo, M. 2009. There is no magic circle, *Games and Culture*, 4(4), 408–417. https://doi.org/10.1177/1555412009343575

Crawford, G. 2015. Is it in the game? Reconsidering play spaces, game definitions, theming, and sports videogames. *Games and Culture*, 10(6), 571–592. https://doi.org/10.1177/1555412014566235

Csikszentmihalyi, M. 2002. *Flow: The Classic Work on How to Achieve Happiness*, revised ed., New York, NY, Rider.

Fishman, J. M. 2017. *Death Makes the News: How The Media Censor and Display the Dead*, New York, NY, New York University Press.

FromSoftware. 2011. *Dark Souls*, Tokyo, Bandai Namco.

FromSoftware. 2014. *Dark Souls II*, Tokyo, Bandai Namco.

FromSoftware. 2016. *Dark Souls III*, Tokyo, Bandai Namco.

FromSoftware. 2018. *Dark Souls Remastered*, Tokyo, Bandai Namco.

Giant Sparrow. 2017. *What Remains of Edith Finch*, Annapurna Interactive.

Guerrilla Games. 2017. *Horizon Zero Dawn*, Sony Interactive Entertainment.

Huizinga, J. 1949. *Homo Ludens: A Study of the Play-Element in Culture*. London, Routledge & Keegan Paul.

Linser, F., Ree-Lindstad, N. and Vold, T. 2008. The magic circle: game design principles and online role-play simulations. Paper presented at ED-MEDIA 2008, Vienna, Austria.

McAllister, K. S. and Ruggill, J. E. 2018. Playing to death, *American Journal of Play*, 11(1), 85–103.

Mellor, P. A. and Shilling, C. 1993. Modernity, self-identity and the sequestration of death. *Sociology*, 27(3), 411–431. https://doi.org/10.1177/0038038593027003005

Murakami, H. 2003. *Norwegian Wood*, Translated by J. Rubin, London, Vintage Books.

Numinous Games. 2016. *That Dragon, Cancer*, Numinous Games.

Rogue Factor. 2015. *Mordheim: City of the Damned*, Focus Home Interactive.

Rugg, J. 2000. Defining the place of burial: what makes a cemetery a cemetery?, *Mortality*, 5(3), 259–275.

Salen, K. and Zimmerman, E. 2003. *Rules of Play: Game Design Fundamentals*, Cambridge, MA, MIT Press.

Solomon, S., Greenberg, J. and Pyszczynski, T. 2015. *The worm at the core: On the role of death in life*, London, Penguin.

Stebbins, R. A. 2009. *Leisure and Consumption: Common Ground/Separate Worlds*, Basingstoke, Palgrave Macmillan.

Urbonas, J. 2012. Designing death: G-design, fatal aesthetics and social science fiction. In *The Edge of Our Thinking*, Ed. F. A. Schmidt, pp. 32–46, London, Royal College of Art.

Walter, T. 2019. The pervasive dead, *Mortality*, 24(4), 389–404. https://doi.org/10.1080/13576275.2017.1415317

Young, C. and Light, D. 2016. Interrogating spaces of and for the dead as 'alternative space': cemeteries, corpses and sites of dark tourism, *International Review of Social Research*, 6(2), 61–72. https://doi.org/10.1515/irsr-2016-0009

Part 1

Playing with Understandings

Chapter 1

Nonhuman Games: Playing in the Post-Anthropocene

Paolo Ruffino

Abstract

This chapter explores what video games can teach us in light of the ongoing sixth mass extinction in the history of our planet, allegedly caused by global warming and the over-consumption of vital resources. Games made and played by nonhuman actors can shed light on the situatedness and partiality of our knowledge regarding the boundaries that separate and differentiate human and nonhuman, interactivity and passivity, entertainment and boredom, and life and death. Nonhuman games help us to articulate the space and time in-between these dualisms and have the potential to reroute gaming (and game studies) from false myths of agency, interactivity, and instrumentalism, and the masculinism inherent in these notions. Nonhuman games are companions for earthly survival, and as such they can be taken as useful references when considering a more ethical approach to the ecological crisis of the Anthropocene. The chapter investigates notions of posthumanism, interpassivity, and contemporary critiques of the early assumptions of game studies on the agency of human players. It looks at video games that play by themselves, idle and incremental games, and the emergence of nonplaying characters in ludic and open-world simulations. It explores forms of automatic play and the use of bots and Artificial Intelligence (AI) in online role-playing games, procedurally generated virtual environments, and games that far exceed the lifespan of their players.

Keywords: Anthropocene; sixth mass extinction; nonhuman; posthuman; agency; interpassivity

Death, Culture & Leisure: Playing Dead, 11–25

doi:10.1108/978-1-83909-037-020201008

Let's Start from the End

Animal species across the planet are extinguishing at a much faster pace than what would normally be expected in a transitionary period of biodiversity loss (Ceballos et al., 2017), an article published in the *Proceedings of the National Academy of Sciences* reported. This publication confirmed the outcomes of previous studies on the alleged sixth mass extinction (Kendall, 1992; Ripple et al., 2017). The causes have been identified in global warming: the melting of permafrost and the rise of temperatures across the globe have been affecting the life of numerous species. The over-consumption of resources by the human population, which in its own turn is raising to an expected figure of 10 billion (Dorling, 2013), has been catastrophic for biodiversity. Guy McPherson, a scientist from University of Arizona, has been claiming throughout his career that the effects of human intervention are now irreversible, and we should expect life to disappear from Earth by 2030 (Curry, 2013). The prediction would be consistent with the outcomes of the previous mass extinctions that our planet has seen in its history, which annihilated 75%–95% of biological forms in each round. McPherson's predictions have still not manifested in their full extent, but he has become a popular reference in the age of social media when his interviews were published on YouTube: a sign of the growing interest and preoccupation for global warming. In a shorter term, a recent United Nations report has confirmed that up to a million species are at risk of disappearing within the next decade (United Nations, 2019), a pace of extinction that is 'tens to hundreds times higher than it has been, on average, over the last 10 million years' (BBC, 2019).

These are only some of the implications of the Anthropocene, the geological epoch characterised by the long-lasting impact of the human species on Earth. Theorised by scientists Eugene Stoermer and Paul Crutzen in early 2000, the Anthropocene is now a widely accepted term to identify the geological impact that we are both causing and witnessing as a species. The Anthropocene has gathered the attention of scholars from different fields, extending far beyond geology and biology. The gesture of naming the present time as defined by our own presence is anthropocentric, but it has the contradictory effect of putting our species at the centre of its own failure. It is self-congratulatory (we have made an impact that will last after our departure) and humiliating (we might not last to see those effects). It attributes *potestas* (power) to humankind, the power to subjugate life and matter, while denying *potentia* to alter its own destiny and generate new futures. Within the same gesture, it acknowledges that human beings have the capacity to inscribe on the surface of planet Earth long-lasting signs of their presence, and it deprives our species (particularly in its most apocalyptic versions) of any significant agency. If there is an Anthropocene in the history of our planet, then there must also be a post-Anthropocene, in which humans will no longer play a significant role or might be completely absent.

This chapter explores how video games could help us re-think the meaning and value of our time on planet Earth, soliciting an ecological thought and a different understanding of the conflictual and contradictory condition prefigured by the notion of the Anthropocene. The chapter will look at how some recent video games have complicated the relation between a gamer and a game, decentring the human and occasionally making humanity completely redundant. At the same time, these ludic texts force us to rethink the boundaries of our self and how we participate in our surroundings, bringing to the fore new perspectives on the meaning of entertainment and boredom, interactivity and passivity, and life and death. Posthuman thought can help us understand how to play these games, as they shift the boundary between our bodies and the machine, the human and non-human (Braidotti, 2013). It is worth exploring these games as they provide alternative modes of thinking about the meaning of being human, and offer insights on how to live better in the time we have left. *Better* does not necessarily mean that they can provide a solution, a fixing, to the wrongdoings of humankind, but they could suggest different ways of enjoying the present time. As I would like to argue, these video games might appear minimal and almost insignificant in light of the catastrophic effects of the Anthropocene. However, it is their marginality within the larger picture that makes them useful and relevant for our struggles.

Boys' Tools

Discourses around the Anthropocene often include a technological solution. The Canadian company Carbon Engineering (https://carbonengineering.com), for instance, has been promoting a carbon removal technology that 'hoovers' carbon dioxide (CO_2) from the atmosphere, and could potentially decrease the amount of carbon emission in a relatively short time. Welcomed as an immediately effective fix to climate change, it exemplifies the conservativism inherent in the technological solutionism that lies at the root of the problem. As such, it does not solve the systemic contradictions that caused global warming in the first place. More specifically, the tools invented by Carbon Engineering include the Air to Fuels technology, which transforms CO_2 in synthetic fuels to be used on already existing cars, aeroplanes, and other transportation vehicles. It is a *patch*, to borrow a term from software engineering, that can be installed without changing the available infrastructure and the industry that supports it.[1] In this view, neoliberal capitalism can continue in its friction-less growth as it repairs its own mistakes. More problematically, the application of Carbon Engineering's large scale hoovers (known as Direct Air Capture technology) is expected to work on a planetary scale, but its effects have been based so far on computer simulations. As argued by Anderson and Peters (2016), relying on these predictions without first seeing

[1]This 'patch' is a metaphor borrowed from the semantic field of informatics rather than tailoring. A tailored patch would be less problematic in this respect. Mending a hole requires awareness that the same patch might not be equally suitable for every piece of clothing. It also implies that fabric does not appear out of nowhere, it must be cut from another piece, inevitably opening a new hole. The gendered connotations of the two practices (software engineering and tailoring) indicate two different modalities of knowledge about ecological crises and their solutions. In this respect, we need feminist thinking more than software updates.

the results of its applications would be a 'high-stakes gamble', which risks locking the planet in a high-temperature pathway if it is not immediately verified. As the authors observe, the presence of these prototypes has already produced the negative effect of delaying political action, as their efficacy has already been taken for granted despite being so far only theoretical.

The Anthropocene and its solutions share a masculinist excess of confidence that aims to control both the present and the future by betting on the effects of new products and technologies. In this view, the ecological crisis is resolved recurring to scalable solutions: a CO_2 hoover produced in Canada should be copy-pasted in other regions of the world and is expected to reproduce the same results everywhere it is applied (around 10,000 of these need to be installed to eliminate CO_2, Carbon Engineering has estimated). The imaginary that sees the Anthropocene as being the same everywhere, resolvable with the same *trick*, and equally affecting every living species, serves to reduce particularisms and frictions. The same approach is visible in the news articles on climate change that take great visibility on social media feeds. These often include in their title a year of no-return, an ultimate deadline after which life will be obliterated from Earth. It is easy to imagine that similar 'expiry dates' will not work as imagined in the narratives of these press releases, even in the most apocalyptic scenario. The disappearance of life from Earth is more likely to be made of molecular events, many of which are already happening, despite being almost invisible. The disappearance of bees and the melting of permafrost are slow and nonspectacular effects of global warming that do not occur at a precise place and time. The Syrian political crisis of 2011 that brought more than 6 million people to migrate is primarily an ecological crisis, as it was induced by the desertification of the region which made access to freshwater harder for the inhabitants (Gleick, 2014). The Anthropocene highlights the numerous frictions and inequalities that take place around, over, and below the surface of the planet, and how these are often out of reach of our limited and situated experience.

Technological solutionism and its inherent masculinism are too confident in their own possibilities. As Joanna Zylinska (2014, 77–88) argues, in these times of crisis we need a minimal ethics for the Anthropocene: a philosophy of life that makes of its vagueness and openness a critical premise. Embracing this mode of thinking involves, as she puts it via the work of Darin Barney (2013), a post-masculinist courage: the courage to face uncertainty and the unpredictable without trying to control it with instrumentalism and rationality – two concepts that video game culture knows a great deal about.

Posthuman Games

The connections between the Anthropocene and video game culture are manifold. Instrumentalism, rationality, and masculinism take a dominant position in both discursive fields. Making and playing video games, as much as naming and trying to fix the Anthropocene, are typically identified as male activities (Shaw, 2011, 2014; Paul, 2018). Game design is often thought of as the art of framing meaningful problems and their solutions (Salen and Zimmerman, 2003). From a historical and ideological perspective, video games are born out of the cybernetic thinking that is at the root of the imagined detachment of mind and body, user and machine, and culture and nature (Crogan, 2011; Pias, 2017). Game studies have been taking varied positions in this respect, occasionally maintaining these dualisms and other times trying to challenge their theoretical and political implications.

At the centre of both game studies and ecological thinking, there is the problematisation of the correlation between humans and the world surrounding them. Notions of agency and interactivity have largely shaped the early debates in study of games. Interactivity has been identified as the defining characteristic that distinguishes games from other forms of expression. However, its vagueness and implied determinism induced rejections and redefinitions. Lev Manovich (2001, 71) claims that interactivity is too broad and vague as a concept, and often associated with marketing jargon, to be of any use for a theoretical understanding of new media. On the other hand, he acknowledges that digital games are 'explicitly participational'. Janet Murray (1997, 2011) prefers the term 'agency' to identify the key design value of digital artefacts. Espen Aarseth (1997, 51) argues that interactivity is 'a purely ideological term, projecting an unfocused fantasy rather than a concept of any analytical significance', although he maintains that 'nontrivial effort is required to allow the reader to traverse the ergodic text' (Aarseth, 1997, 1). Interaction has been re-defined and renamed. To be addressed from a theoretical standpoint, it required to be cleared of the overtly-enthusiastic and deterministic claims that put it at the heart of the digital revolution. However, the question on how to properly evaluate the relation between user and machine reinforced the presence of the dualism, even when the two terms were written 'under erasure'. When not explicitly brought into question, the implied separation between user and system has allowed numerous authors to think of games as texts made of rules and problems to be understood, resolved, and fixed (Salen and Zimmerman, 2003; Juul, 2005). These visions have reinforced an imagined separation (both spatial and cognitive) between players and their games.

Posthuman thought has served to deconstruct the dualism between an interactor and a machine, shedding light on the lived assemblage of human and nonhuman minds, bodies, and environments, and imagining alternative modes of thinking. Seth Giddings (2005) argues that a theorisation of human and nonhuman agency is pivotal in the study of digital games. Games aestheticise and realise Donna Haraway's (1991) figure of the cyborg, a coming together of organic and technological that is intimate, unknowable, ironic, and monstrous. When playing a video game, Giddings (2005) observes the flow of information does not originate from the human or machine but takes place within the cybernetic loop that makes the two sides indistinguishable and no longer separable. Alexander Galloway (2006) argues that the two sides, 'operator' and 'machine', are inevitably drawn together in the act of playing a video game. Operator and machine's actions take place at the diegetic and nondiegetic levels of the game, manipulating its narrative and internal functions (as it happens when 'saving' a game, for instance). More recently, Brendan Keogh has observed that a video game text is produced through the co-presence of the player's body and the hardware, generating playful encounters between the two (Keogh, 2018, 47). The imaginary of games as being played by a disembodied player from an empty position is critiqued by Keogh, as projecting the fantasy of a text made of pure information and mechanical interactions. The presence of the player's body and the actions taken while playing a video game happen with and through a nonhuman agent. The possibilities of a video game are generated by mobilising the physicality of human and nonhuman actors, and by destabilising the definitions of both. Video games are, even from a narrative and semiotic perspective, texts of unstable boundaries (Fassone, 2017). A posthumanist understanding of the medium of the video game complicates notions of interactivity and agency. In so doing, it highlights how player's actions are more similar to a bet on the future, a 'pre-emptive control and virtual, remote regulation of contingency' (Crogan, 2011, 32). Game actions are motivated by the uncertainties generated by the 'collusion' (to borrow a term from Giddings, 2009) of human player and nonhuman machine. When playing games' identities are fractured, 'contradictory, partial, and strategic' (Haraway, 1991, 155).

As argued by Katherine Hayles (1999, 21), the process through which cybernetics has detached information from its body has brought with it the assumption that stability is a desirable goal, and that 'human beings and human social organizations are self-organizing structures'. The failure of these assumptions is, I argue, at the origin of many delusions and misunderstandings around the current ecological crisis. As Timothy Morton (2017, 14) observes, the difficulty we are currently experiencing when thinking about the Anthropocene is rooted in the 'severing' between reality and real, which has created an 'impermeable human world'. In this all-too-human bubble, we are prevented from conceiving solidarity with nonhumans, persuaded by the idea that human culture is the primary filter that makes things real. Morton's ecological thought proposes to multiply the number of differences in our understanding of what constitutes humankind and to focus on how bodies are entangled in the surrounding environment and composed of a multitude of biological organisms that outlive the 'human' as a singular entity. For Morton (2017), inverting a common-sense statement, the whole is less than the sum of its parts. Similarly, Anna Tsing notes that modern knowledge tends to 'sum up' and reduce particularity. By contrast, attending to life's 'contaminated diversity' makes various forms of collaborative survival come to light (Tsing, 2015). Playing video games, posthumanist thought observes, depends upon the opening up of human boundaries. Bodies, gestures, thoughts, and identities are made porous, rendered unstable, and the principle of a self-organising structure is lost in the act of playing. The genealogy that connects video games with cybernetics keeps two opposing imaginaries together: on the one hand, a separation between human and nonhuman, which sees human actors solving problems from an invisible position, while on the other there is a collaboration, a physical and intellectual melting of human and nonhuman that makes the separation between the two no longer relevant.

This chapter argues that video games tend to destabilise the human, and that instrumentalism and rationality are challenged particularly through contemporary forms of digital gaming. The connection between ecology and video games might not appear obvious, but it lies in the detachment of bodies and information, human and nonhuman actors, that frames cybernetics and the culture surrounding it. However, as argued by posthumanist perspectives on game studies, video games are contexts where human and nonhuman actors think, act, and play together in much more complex manners. This line of thinking can bring us back to the ecological and ethical question of how to live a better life in the time we have left on this planet.

Posthumous Games

In his analysis of game actions, Galloway (2006) takes into account the game *Shenmue* (SEGA AM2, 1999) by Yu Suzuki. *Shenmue* was published by Sega on the console Dreamcast, and was one of the first open world video games of the era. The story of the game begins in Yokosuka, Japan, in November 1986, where the main character, Ryo Hazuki investigates the murder of his father. While engaging with the local villagers and collecting evidence and clues that can lead him to the perpetrators of the crime, the player immerses him or herself in a living environment, where nonplaying characters follow precise routines and patterns. An internal clock regulates day and night cycles, shops have opening times, while some key characters can only be approached at a specific time and place. When the player does not interact with the simulation, leaving the controller and letting the game rest, the fictional representation of Yokosuka will continue running: characters can be seen walking around the streets, carrying on with their own businesses. This type of open-world simulation is now considered a standard in the genre of adventure games. In games such as *My Time at Portia* (Pathea, 2018), nonplaying characters populate a post-apocalyptic world and follow a series of time-based patterns. Recent titles such as *Red Dead Redemption 2* (Rockstar San Diego, 2018), *Fallout 76* (Bethesda Game Studios, 2018), *Assassin's Creed Odyssey* (Ubisoft Quebec, 2018) are further instantiations of this approach to game design, where computational power of contemporary consoles and PCs is often deployed to increase the number of characters, dialogues, events, and graphical details that are simulated as if they were living a life of their own, activating themselves without the input of the player.

As Galloway (2006, 18) observes, while playing *Shenmue* the operator can step back and 'look at the poetry of the algorithm'. When the human player decides not to intervene, the correlation between human and nonhuman is no longer balanced. One side takes the lead:

> When games like *Shenmue* are left alone, they often settle into a moment of equilibrium. Not a tape loop, or a skipped groove, but a state of rest. The game is slowly walking in place, shifting from side to side and back again to the center. It is running, playing itself, perhaps. The game is in an ambient state, an *ambience act*. (Galloway, 2006, 10)

Despite functioning through a cybernetic relationship, the game can also live in an independent manner. Machine actions might appear to be self-sufficient, almost prefiguring the possibility of a game played without human's intervention.

This almost paradoxical modality of play alludes to the fears of the post-Anthropocene: a world where human beings are no longer central in the functioning of the planet Earth or are perhaps entirely extinguished. Narratives around the Anthropocene implicitly suggest that our time spent on this planet is limited, and our presence destined to end, as all the other major forces that have shaped previous epochs. A future Earth without humans is, in other words, possible, and a video game that plays by itself helps us to imagine this unimaginable scenario.

If video games can complicate our understanding of our involvement in the environment surrounding us, then video games that completely erase human's presence are *things to think with* when elaborating the imagination of life after the human. They are the playful equivalent of the television show Life After People, broadcast by History channel in 2008 (as a 2-hour special) and 2009 (as a series). The latter show visualises what would happen to animals, artefacts, cities, buildings, and landscapes in the event of human disappearance. The reasons for the absence of human beings are not explored. The show is intended as a purely fictional work of scientific imagination, although some of the locations shown are actual examples of places that have been abandoned by human population for various reasons. The fictional Yokosuka in *Shenmue* lives a life of its own, and it is a life *after people*: no longer dependent on the inputs of a human player in flesh and bones. It is similar to the not distant area of Fukushima, also in Japan, which could have appeared in the History show if it was shot today. Rendered inhabitable in 2011 after a tsunami hit the Daichii nuclear power plant, the area is now slowly being repopulated with animals and vegetations. While the radioactivity level makes life dangerous for human beings in the long run, the absence of humans is advantageous for forms of life that were previously marginalised because of man-made pollution and constructions. The Fukushima nuclear area is becoming what science fiction author Bruce Sterling has defined an 'involuntary park': areas that are no longer useful for capitalistic exploitation, because they are too dangerous or polluted, are abandoned, and then slowly reoccupied by nonhuman forms of life (SCI-Arc Channel, 2018).[2] Games such as *The Last of Us* (Naughty Dog, 2013) are played in similar settings, once populated by humans and then rapidly transformed into hostile environments, but still prolific of nonhuman or barely-human forms of life.

Video games are premediating the post-Anthropocene in numerous ways. Björk and Juul (2012) have explored the theoretical implications of 'zero player games', where the human player is eliminated or radically marginalised in the game mechanic. The games discussed in the next session are made or played by nonhuman actors, but the human is only temporarily removed, as if they were 'involuntary parks' deprived of human presence.

---

[2]A similar phenomenon is taking place in Chernobyl, Ukraine, where the territory abandoned after the nuclear disaster of April 1986 (a few months before the beginning of *Shenmue*, coincidentally) is now considered a healthier environment than most European cities (Smith, 2007).

Idle Games

The genre of idle games is a growing trend in the digital entertainment industry. It is a type of game which is 'left running with little player interaction' (Alharthi et al., 2018, 1). The ambience act discussed by Galloway in relation to *Shenmue* is almost transformed in a game of its own. These titles are also known as background, ambient, incremental or clicker games. Most of the idle games are 'played' on smartphones, although the most popular titles are also published on consoles such as PlayStation 4 and internet browsers. In principle, these games only require to be activated and then carry on independently, although they typically involve, or encourage, a minimal form of interaction, such as repeated clicking. The game *AdVenture Capitalist* (Hyper Hippo Productions, 2014) is a 'capitalism simulator', where the production of profit can be automatised by the player in the early stages of the game. When returning to the app, even after months or years from its activation, the capital will have automatically raised. The wealth accumulated can be spent on businesses and upgrades (such as 'lemon stands', 'oil companies', and various types of managers) that increase the pace of capital accumulation. There is no final goal or objective in this fictional capitalistic endeavour. The total profits can rapidly escalate to impossible figures such as undecillion or vigintillion dollars, and towards theoretically infinite numbers. Idle games are often ironic and acknowledge their paradoxical design. The popular *Progress Quest* (Fredericksen, 2002) is a parody of the more famous *EverQuest* (Verant Interactive and 989 Studios, 1999), an online role-playing game. *Progress Quest* preserves the level-up feature typical of the design of Role-Playing Games (RPG) but eliminates the human labour involved in the process. The tasks are automatised, and the player is only expected to watch numbers increase on screen. The game *Cookie Clicker* (Thiennot, 2013) lets players choose from a series of bizarre upgrades that increment the baking of cookies at exponentially higher rates. Once the game is closed, it continues to play itself, detached from human vision. When the player returns to the game, they will find that the score has increased over time.

Contemporary idle games have transitioned from experimental parodies into a proper genre. Their core mechanic of play has been normalised through commercially successful productions such as *AdVenture Capitalist*, *Clicker Heroes* (Playsaurus, 2014), and *Idle Heroes* (Droidhang Games, 2016; Deterding, 2016). Idle games are part of the broader phenomenon of casual games: products released for an audience of smartphone and laptop owners, who would not normally commit to a demanding experience and consider themselves 'unable to fit a game into their lives' (Juul, 2010, 5). As argued by Keogh and Richardson (2018), while these games do not require an intensive form of engagement, players occasionally decide to spend a significant amount of time in planning, clicking, and elaborating strategies to maximise their (minimal) effort. Micro-transactions are often included in idle games, and a significant portion of players spend real money to receive upgrades and aesthetic ameliorations. Idle games might be independent of human players, and theoretically self-functioning, but they still involve the player intellectually or economically via micro-transactions. The contemplation of the ambience act, to put it through Galloway's terminology, can be demanding and prolific of diverse forms of engagement.

While analysing idle games, Sonia Fizek has noticed how they pose a serious challenge for game studies, as they complicate the notion that enjoyment derives from the active engagement with the ludic text. She argues that 'idling may be understood as delegated pleasure derived from the act of outsourcing gameplay onto the game itself' (Fizek, 2018, 142). Idle games are an example of interpassive experience: neither interactive nor passive, these games 'provide their own reception' (Pfaller, 2017). However, it matters what kind of interpassive experience is presented. While some idle games are more experimental, others tend to exploit the user's addiction via microtransactions.

Other titles instead try to provoke more complex imaginaries. *Universal Paperclips* (Lantz, 2017) is similar in its mechanic to other incremental and idle games. The game simulates a paperclip factory that exponentially increases in size and automatises its production through an artificial intelligence. The AI will not stop until the entire matter of the universe has been transformed into paperclips. The exponential and unlimited growth of idle games is here confronted with the imaginary of friction-less capitalism that underpins it. The game reminds that matter produces frictions, and that the nonhuman has its own material limits. It also questions the risks associated with the appeals of automatisation and delegation of labour. The fictional Artificial Intelligence governing the production of paperclips cannot be stopped and will determine the extinction of human beings, suggesting that similar experiments in automation within the military-industrial complex might escape our control and produce catastrophic effects. Looking at an earlier example, *Cow Clicker* by Ian Bogost (2010b) has been defined as an inspiration for the genre of 'waiting' games (Tanz, 2011). The game is playable via Facebook and replicates the strategy of clicking and waiting that is at the core of games such as *FarmVille* (Zynga, 2009). While intended as a parody of those games, Bogost's *Cow Clicker* (2010b) became an unexpected success among casual players who were not necessarily aware of the critical intent. Bogost has developed this line of critical design and released games that are entirely based on waiting. The series *A Slow Year: Game Poems*, produced in 2010 for the 1977 Atari VCS console, is comprised of four games, one for each season, and each requiring 'a different kind of sedate observation and methodical input', rather than 'action or strategy' (Bogost, 2010a). The earlier *Guru Meditation* (Bogost, 2009) is designed to be played while meditating on the *joyboard*, an unusual interface produced by Amiga in 1982, where players could sit or use their feet to interact with a game. Both games function only to the extent that the player disappears from the scenario, by standing still, meditating, repeating the same minimal actions. At the same time, these games invite their players to reflect on their own presence, the duration of the experience, and the relation established with the hardware.

Delegated Games

The act of play can become a delegated activity: a spectacle to watch, for instance on Twitch or YouTube, or to be completely emancipated from. Stefano De Paoli (2013) has investigated how the repetitiveness and boredom derived from levelling up in Massive Multiplayer Online Role-Playing Games (MMORPGs) prompts players to find alternative solutions to their laborious tasks. The automation of these activities via bots and macros, which De Paoli refers to as 'automatic-play', introduces a critique of the in-game economic systems and their representations of labour. While automatic play often takes place at the boundaries of game rules, and is not always tolerated by developers and publishers, it has also prompted game companies to produce titles intentionally designed for this type of activity. *Screeps* (2016) is an open-source MMO (massive multiplayer online) game for programmers. The main mechanic of the game consists in controlling the artificial intelligence of the player's units via JavaScript. Units compete against those of other players 24 hours a day, even when players are offline. The game has become popular among programmers, up to a point that it is taken into account when recruiting software engineers.

A similar approach has been embraced by the popular series *Forza* (Xbox Game Studios, 2005–2019). Published by Microsoft on its *Xbox* console, the *Forza* series has been promoting the *Drivatar* system as a key component (Bittanti, 2015). The driving game lets players train an artificial intelligence that simulates the driving style of the player. When the game is turned off, a *drivatar* will replace the real player and compete in online tournaments, earning in-game currency and winning trophies. The *drivatar* expands on the concept of the 'ghost car', already present in driving simulations, where a ghostly representation of the fastest performance remains visible on track to provide a reference to the driver who attempts to beat the current record. Both ghost cars and *drivatars* open paradoxical and *creepy* scenarios. Online forums frequently report stories of players who pass away and leave their virtual representations behind, still playing the game on their consoles. While ghost cars of dead players can be easily found in saved games of abandoned consoles, the *drivatar* appears in online games and competes with other players who might not be aware that they are being challenged by a dead person. In the long run, if the servers that host the *Forza* games will remain functioning for a sufficiently long time, the entire game will be played by *drivatars* of dead humans.

Procedural Games

The phantasy of a game that exceeds the human is epitomised by *No Man's Sky* (Hello Games, 2016). The game has been released after a controversial crowdfunding campaign. The relatively small team announced that they had developed a software for the procedural generation of 3D environments, which would allow them to produce a game far exceeding the possibilities offered by any other title. The game is set in a fictional online universe with animals, vegetation, and resources to be explored and collected by the players. It is estimated that it would take 585 million years to visit every single planet in the game, a time that exceeds the known age of our universe. Hello Games announced in the development stage that artificial bots have been used to check the consistency of the universe, as it would have been impossible for a human being to test such a gigantic environment. Thus, *No Man's Sky* is mostly made of areas produced by a nonhuman agent (the software used to procedurally generate the environments) which have been only explored by nonhuman agents (the AI used to test the game).

No Man's Sky cannot be completed within the Anthropocene. In fact, it cannot be completed within any geological age. To be more precise, it is not a game to be completed, or intended to be played, by any human being. Not only because of its magnitude, but because of its inhumane boredom (a *feature* that many consumers vehemently complained about after its release). The procedurally generated universe is immense, but also invariably uneventful and filled with planets that look very similar to each other. The players' disappointment was justified by the hype generated through the crowdfunding campaign. YouTube and Reddit have been the privileged platforms where to expose the infinite repetitiveness of the game. The imaginary of a nonhuman game, produced by a nonhuman agent and framed on nonhuman times, has turned against the developers in an all too human manner. The criticism brought gamers to send death threats to the development team, accused of having betrayed their supporters (MacDonald, 2018).

Conclusion: Playing the Anthropocene

This chapter has investigated the paradoxes and contradictions, as well as the irony, *weirdness,* and *creepiness*, of video games made or played by nonhuman agents. As Haraway's vision of the cyborg did with cybernetics, nonhuman games confuse and complicate the ontologies of digital texts, and shed light on the situatedness, temporality, and partiality of our knowledge (Haraway, 1991; Kember, 2017). Life might be disappearing from Earth at some point, but we are not there yet. We are in-between birth and death, the beginning and the end, and we have always been. Nonhuman gaming helps us to articulate this space and time in-between and has the potential to re-route gaming (and game studies) from false myths of agency, interactivity, and instrumentalism, and the masculinism inherent in these notions. Nonhuman games are companions for earthly survival, and as such they can be taken as useful references when considering a more ethical approach to the ecological crisis of the Anthropocene. While thinking of climate change as an irremediable global issue might pre-empt any form of political decision, considering instead the complexities of our situated encounters with the nonhuman, even in the games we (do not) play, suggests a more humble and attentive approach.

Interactions and Sensoria

Cheng, I. 2015–2017. *Emissaries Trilogy*.

Kenji Kubota, E., Mattes, F. and Waite, J. (Curators). 2015. *Don't Follow the Wind* [Group exhibition].

Mattes, E. and Mattes, F. 2016. *Fukushima Texture Pack*.

Molleindustria. 2019. *Lichenia*.

Watanabe, B. 2016. *San Andreas Deer Cam*.

References

Aarseth, E.J. 1997. *Cybertext: Perspective on Ergodic Literature*, Baltimore, MD, The Johns Hopkins University Press.

Alharthi, S.A., Alsaedi, O., Toups, Z.O., Tanenbaum, J. and Hammer, J. 2018. Playing to wait: a taxonomy of idle games. In *Proceedings of CHI 2018*, 21–26 April, pp. 1–15, Montreal, Canada.

Anderson, K. and Peters, G. 2016. The trouble with negative emissions. *Science*, *354*(6309), 182–183.

Barney, D. 2013. Eat your vegetables: courage and the possibility of politics, *Theory & Event*, 14(2), 46942. https://doi.org/10.1353/tae.2011.0023

BBC. 2019. UN says one million species face extinction. *BBC*, 7 May. Available at: https://bbc.in/2qSTOkJ

Bethesda Game Studios. 2018. *Fallout 76*. Bethesda Softworks.

Bittanti, M. 2015. *Orizzonti di forza: Fenomenologia della guida videoludica*, Milan, Edizioni Unicopli.

Björk, S. and Jesper, J. 2012. Zero-player games, or: what we talk about when we talk about players. In *Proceedings of the Philosophy of Computer Games Conference Madrid 2012*, Madrid. Available at: https://www.jesperjuul.net/text/zeroplayergames/

Bogost, I. 2009. *Guru meditation* [video game].

Bogost, I. 2010a. A slow year: game poems, *A Slow Year: A Chapbook of Game Poems for Atari VCS, PC and Mac*, Available at: http://bit.ly/2MSPytR

Bogost, I. 2010b. *Cow Clicker* [video game].

Braidotti, R. 2013. *The Posthuman*, Cambridge, Polity Press.

Ceballos, G., Ehrlich, P.R. and Dirzo, R. 2017. Biological annihilation via the ongoing sixth mass extinction signalled by vertebrate population losses and declines, *Proceedings of the National Academy of Sciences*, 114(30), E6089–E6096. https://doi.org/10.1073/pnas.1704949114

Crogan, P. 2011. *Gameplay mode: War, simulation and technoculture*, Minneapolis, MN, University of Minnesota Press.

Curry, N. 2013. Some credible scientists believe humanity is getting verrrrrrry close to extinction, *Vice*, 22 August. Available at: http://bit.ly/2NmNF7G

De Paoli, S. 2013. Automatic-play and player deskilling in MMORPGs, *Game Studies*, 13(1). Available at http://gamestudies.org/1301/articles/depaoli_automatic_play

Deterding, S. 2016. Progress wars: idle games and the demarcation of "real" games. In *Proceedings of 1st International Joint Conference of DiGRA and FDG*, 1–6 August, Dundee, Scotland. Available at: http://www.digra.org/wp-content/uploads/digital-library/paper_267.pdf

Dorling, D. 2013. *Population 10 billion: The coming demographic crisis and how to survive it*, London, Allen Lane.

Droidhang Games. 2016. *Idle Heroes* [video game].

Fassone, R. 2017. *Every Game Is an Island: Endings and Extremities in Video Games*, London, Bloomsbury.

Fizek, S. 2018. Interpassivity and the joy of delegated play in idle games. *Transactions of the Digital Games Research Association*, 3(3). 137–163. https://doi.org/10.26503/todigra.v3i3.81

Fredericksen, E. 2002. *Progress Quest* [video game].

Galloway, A. 2006. Gaming: Essays on Algorithmic Culture. Minneapolis (MN): University of Minnesota Press.

Giddings, S. 2005. Playing with non-humans: digital games as techno-cultural form. In *Proceedings of DiGRA 2005 Conference*, 16–20 June, Vancouver, Canada. Available at: http://www.digra.org/wp-content/uploads/digital-library/06278.24323.pdf

Giddings, S. 2009. Events and collusions: a glossary for the microethnography of video game play, *Games and Culture*, 4(2), 144–157. https://doi.org/10.1177%2F1555412008325485

Gleick, P.H. 2014. Water, drought, climate change, and conflict in Syria, *American Metereological Society*, 6(3), 331–340. https://doi.org/10.1175/WCAS-D-13-00059.1

Hayles, N. K. 1999. How We Became Posthuman: Virtual Bodies in Cybernetics, Literature, and Informatics. Chicago and London:The University of Chicago Press.
Haraway, D. 1991. A cyborg manifesto: science, technology, and socialist feminism in the late twentieth century. In *Simians, Cyborgs and Women: The Reinvention of Nature*, Ed. D. Haraway, 2nd ed., 149–181. London, Free Association Books.
Hello Games. 2016. *No Man's Sky*, Hello Games.
Hyper Hippo Productions. 2014. *AdVenture Capitalist*, Kongregate.
Juul, J. 2005. *Half-real: Video Games between Real Rules and Fictional Worlds*, Cambridge, MA, MIT Press.
Juul, J. 2010. The Casual Revolution: Reinventing Video Games and their Players. Cambridge (MA): The MIT Press.
Kember, S. 2017. After the anthropocene: the Photographic for earthly survival?, *Digital Creativity*, 28(4), 348–353. https://doi.org/10.1080/14626268.2017.1380048
Kendall, H.W. 1992. *World Scientists' Warning to Humanity*, Cambridge, The Union of Concerned Scientists.
Keogh, B. 2018. *A Play of Bodies: How We Perceive Videogames*, Cambridge, MIT Press.
Keogh, B. and Richardson, I. 2018. Waiting to play: The labour of background games. *European Journal of Cultural Studies*, 21(1), 13–25. https://doi.org/10.1177/1367549417705603
Lantz, F. 2017. *Universal Paperclips* [computer game].
MacDonald, K. 2018. No man's sky developer Sean Murray: 'It was as bad as things can get'. *The Guardian*, 20 July. Available at: http://bit.ly/2MSbrcR
Manovich, L. 2001. *The Language of New Media*. Cambridge, MIT Press.
Morton, T. 2017. *Humankind: Solidarity with Nonhuman People*, London, Verso.
Murray, J.H. 1997. *Hamlet on the Holodeck: The Future of Narrative in Cyberspace*, Cambridge, MA, MIT Press.
Murray, J.H. 2011. *Inventing the Medium: Principles of Interaction Design as a Cultural Practice*, Cambridge, MA, MIT Press.
Naughty Dog. 2013. *The Last of Us*, Sony Computer Entertainment.
Pathea. 2018. *My Time at Portia*, Team17.
Paul, C.A. 2018. *The Toxic Meritocracy of Video Games: Why Gaming Culture Is the Worst*, Minneapolis, MN, University of Minnesota Press.
Pfaller, R. 2017. *Interpassivity: The Aesthetics of Delegated Enjoyment*, Edinburgh, Edinburgh University Press.
Pias, C. 2017. *Computer Game Worlds*, Chicago, IL, The University of Chicago Press.
Playsaurus. 2014. *Clicker Heroes* [video game].
Ripple, W.J., Wolf, C., Galetti, M., Newsome, T.M., Alamgir, M., Crist, E., Mahmoud, I.M., Laurance, W.F. 2017. World's scientists' warning to humanity: a second notice, *Bioscience*, 67(12), 1026–1028.
Rockstar San Diego. 2018. *Red Dead Redemption 2*, Rockstar Games.
Salen, K. and Zimmermann, E. 2003. *Rules of Play: Game Design Fundamentals*, Cambridge, MA, MIT Press.
SCI-Arc Channel. 2018. Bruce Sterling & Benjamin Bratton in conversation, *YouTube*, 5 November. Available at: http://bit.ly/2Jr8U76
Screeps. 2016. *Screeps* [video game].
SEGA AM2. 1999. *Shenmue*, Sega.
Shaw, A. 2011. Do you identify as a gamer? Gender, race, sexuality, and gamer identity. *New Media & Society*, 14(1), 28–44. https://doi.org/10.1177%2F1461444811410394
Shaw, A. 2014. *Gaming at the Edge: Sexuality and Gender at the Margins of Gamer Culture*, Minneapolis, MN, University of Minnesota Press.
Smith, J.T. 2007. Are passive smoking, air pollution and obesity a greater mortality risk than major radiation incidents?, *BCM Public Health*, 7, 49. https://dx.doi.org/10.1186%2F1471-2458-7-49

Tanz, J. 2011. The The curse of cow clicker: How a cheeky satire became a videogame hit. Wired, 20 December. Available at: https://www.wired.com/2011/12/ff-cowclicker

Thiennot, J. 2013. *Cookie Clicker* [video game].

Tsing, A.L. 2015. *The Mushroom at the End of the World: On the Possibility of Life in Capitalist Ruins*, Princeton, NJ, Princeton University Press.

Ubisoft Quebec. 2018. *Assassin's Creed Odyssey*, Ubisoft.

United Nations. 2019. UN report: nature's dangerous decline 'unprecedented'; species extinction rates 'accelerating'. *Sustainable Development Goals*, 6 May. Available at: http://bit.ly/2JsGZUi

Verant Interactive and 989 Studios. 1999. *EverQuest*, Sony Online Entertainment.

Xbox Game Studios. 2005–2019. *Forza* [series], Xbox Game Studios.

Zylinska, J. 2014. *Minimal Ethics for the Anthropocene*, London, Open Humanities Press.

Zynga. 2009. *FarmVille*. Zynga.

Chapter 2

Peaceful in Death: Encountering Death in the Pokémon Universe

Ashley Darrow

Abstract

Pokémon has proven to be one of the most enduring and successful gaming franchises. Over the last two decades, Pokémon has produced dozens of games, multiple animated shows, and several movies. In this time, video games have found a new footing when it comes to approaching death. Titles like *A Mortician's Tale* (2017) by Laundry Bear Games have opened up digital gaming as a place, where death can be explored with serious thought and a new emotional depth. A dichotomy has emerged between games that take death seriously and ones that lower death to a mere game mechanic. This chapter looks to bridge that growing divide and explore the pop phenomenon that is Pokémon through the same critical lens that has been used to examine games like *A Mortician's Tale*. Ultimately, the Pokémon franchise emerges as capable of thoughtful and provocative conversation about death and dying.

Keywords: Pokémon; death; death studies; game studies; video games; popular culture

Introduction

Pokémon is a global phenomenon. Everything, from breakfast snacks to airplanes, has been inducted into the greater set of Pokémon paratexts. Despite this near-total global saturation, the discourse surrounding death in Pokémon remains decidedly limited. Rather than approach the game on its own terms or as a text in its own right; the discourse at large has biased towards ignoring Pokémon or banishing it to the realm of disposable 'pop culture'. We can, and should,

Death, Culture & Leisure: Playing Dead, 27–39

doi:10.1108/978-1-83909-037-020201009

approach Pokémon with the same discursive tools we use to study games more explicitly about death.

Dying in the Digital World

The human artistic focus on our own mortality didn't disappear upon our departure from the physical and into the realm of the digital. Video games have always had a focus on death. Even in the earliest videoludic title *Spacewar!* (Russel, 1962) and the emergent arcade phenomena with games like *Death Race* (Exidy, 1976) had a focus on the end of life as both a core mechanic and theme. Later games only furthered this trend. The advent of shooters with the release of *Wolfenstein 3D* (ID Software, 1992) and *DOOM* (ID Software, 1993) marked death as the mechanic of choice for video games. The popular conception of death in the videoludic, however, became limited both by the focus on *giving* death and popular discourse centreing on outrage.

These games concern themselves with only part of the conceptualisation of death. Shooters, dungeon crawlers, fighting games, and other modes feature *giving* death a central mechanical conceit. You fight, you shoot, you kill. The play here is largely focussed on *giving* death to others and not the experience of death. Nicolucci (2019, 4) notes the fungible nature of typical player encounters with death: 'Death in gaming is also used as a reset function: a player dies over and over again throughout the game to automatically revive after restart'. Our deaths, as players and player characters, are quick. They are marked only by brief 'game over' screens and are designed to prompt us for quarters and get us to try again. Death here is fast and disposable. 'Enemies' are ground into points and our lives are ground in quarters. It's hard to miss the ingrained capitalist realism here.

The academic and popular conversations around death in gaming tend to revolve around the more violent examples. Writing on video games, Durkin (2003, 46) cited 'Mortal Combat and Duke Nukem' as games that 'feature vivid images of violent deaths'. Muncy (2017) highlights this, writing that 'videogames have never really gotten death. They're *full* of death, of course, but only as a conceit, a shallow reading of a much deeper concept'. The idea here is that popular gaming represents death poorly. That 'everything that was directly lived has receded into a representation' (Debord, 2005). Through this mass consumerism 'death' has been abstracted away from a central and defining aspect of human life and towards being only 'a punishment, a roadblock, a temporary setback, an opportunity … never philosophical' (Muncy, 2017). However, this attitude has shifted in recent years with the rise of death positive gaming.

The 'Death Positive' movement derives from anthropological studies into death *a la* Ernest Becker's *The Denial of Death* and, as a term and movement, was coined by famed mortician Caitlin Doughty in a 2013 tweet (Doughty, 2016). The movement, like Sex Positivity from whence the term specifically derives, seeks to normalise, and understands our relationship with a natural human occurrence. This growing cultural appreciation of the phenomenon of death has not left the world of the videoludic behind.

We are in the midst of something of a mortuary gaming boom. With the whirlwind critical success of *A Mortician's Tale* (Laundry Bear Games, 2017;

see Chapter 11 of this volume) and the massive popularity of the *Dark Souls* franchise (FromSoftware, 2011, 2014, 2016; see Chapter 6 of this volume), games that meditate on various aspects of the nature of death and dying abound. *That Dragon Cancer* (Numinous Games, 2016; see Chapter 12 of this volume) explore the painful and complex emotionality of the death of a child while *What Remains of Edith Finch* (Giant Sparrow, 2017; see Chapter 7 of this volume) asks us about how we want to be remembered and who does the remembering. These titles join a host of others: *Continue?0987654321* (Oda, 2014), the grim *The Missing: J.J. Macfield and the Island of Memories* (White Owls Inc, 2018), and the political *Spec Ops: The Line* (Yager Development, 2012) in an emerging ludic discussion of death. While these titles are much closer to being explicitly about death, dying, and the experiences therein, games that lack this central focus shouldn't be beyond the purview of ludic death studies.

A Mortician's Tale (Laundry Bear Games, 2017) launched what would become the definitive death positive video game. *A Mortician's Tale* tells the story of a young woman named Charlie as she starts out and grows her career as a mortician. The game is a slow-paced and contemplative look at the American funeral industry and our industrialised approaches to death. The game was met with near universal praise upon release and has quickly become a focal point for writers looking to parse death in the digital. Nicolucci (2019, p) sums up the academic appraisal of the game:

> Until now, video games have tended to employ death as a mechanic (e.g. for penalty, challenge or learning purposes) or as a narrative element in the story. By contrast, A Mortician's Tale covers and explores death as a biological, emotional and social event.

This framework has also been dominant in the popular discourse as well.

A blog post on *Funeral Direct* (2017) argues:

> Even though a majority of video games out there involve a lot of meaningless death but still games, as a medium, struggle to understand death in its real weight and complexity.

This is the common refrain across the critical discourse of gaming. The argument is, in summation, video games don't take death seriously. It is not *death* as a human experience, but death as merely so many quarters. This struggle for death's meaning does less to remove death from ludic discourse and more to insert it in every absence. Death abhors a vacuum. Writing on the nature of failure in Pippin Barr's *Let's Play: Ancient Greek Punishment*, Anable (2018, 112) notes that the game is 'Damning us to endless repetitions of failure, Barr denies us even the small satisfaction of completion or virtual death'. Despite a perceived absence of death, games are filled with a contestation for the meaning of death. This conversation around the absence of death in gaming resembles the same dialogue surrounding other media.

The weight and importance given to death has shifted in terms of how our media portrays and represents the dead. News media in the United States, for instance, has moved away from depicting the often graphic realities of death and has moved towards a censored and euphemistic display. Fishman (2017, 225) writes that 'When covering death, photojournalists carefully craft pictures that try to hide the subject, but sometimes they fall short of these goals, and then editors turn to digital techniques'. This highlights not only technological similarities – the pixilated bodies on the nightly news fold readily into the pixilated bodies in older digital games – but also signposts an inability or unwillingness to confront death on its own terms. Games are not as removed from other media as the prevelainent popular discourse would suggest.

Funeral Direct (2017) goes on to say 'We see so much *faux* death in games that use death as an empty stand-in'. This *faux* death Funeral Direct notes in gaming maps onto the same disappearance of death as outlined by Fishman. The interpretive framework here might suggest that the 'deaths' in a more popular games like *Pokémon* appear a world apart from *A Mortician's Tale*. Deaths in *Pokémon* are just perfunctory moments that give the player a moment of inconvenience to spur them on to more advanced play, but this is far from a complete analysis of these games as texts and, indeed, cultural artefacts. While death in gaming does have a mechanical function, reducing the discussion to this removes inherent meaning and abnegates discourse.

A much more nuanced conversation comes into play when we note, as Nicoluccia (2019, 1) has, that 'death is present in most video games as a mechanic or a narrative element'. Unsurprisingly, the vast and multitudinous world of Pokémon has not escaped this mortal discourse. A discourse around death and Game Studies is already being built. Rather than looking inward, to games that self-elect themselves as discourses on death, what about a game that, if anything, has self-elected as the bringer of capitalism disguised as youthful adventure and nostalgia?

Death in the videoludic can't be reduced to simple violence or pure escapism. Durkin's (2003, 43) essay 'Death, Dying, and the Dead in Popular Culture' notes that American culture is at once 'death-denying' and at the same time obsessed with death in its media. We see this reflected in our gaming. The popular discourse about juvenile power fantasies belies a deeper reality and ignores the truth that death is always at the forefront of our cultural expressions – regardless of how explicit it is. The Pokémon franchise has been largely left out of our conversations on death and dying. Indeed, the popular discourse around death in the world of Pokémon mirrors that of the larger games industry.

In the *Handbook of Death and Dying,* Durkin (2003, 48) writes 'Children or adolescents playing violent video games must detach themselves from the depictions of primal carnage occurring before their eyes'. This seems to copy over some prevalent, negative, attitudes about play and death. The focus on 'adolescents' infantilises the capabilities of gaming and refuses to meet gaming on its own terms as an art form. Also, this takes the spectacle for all that is to be seen. Further, rather than detach, what happens when we attach ourselves to the videoludic death as presented in the Pokémon games?

The detritus of popular culture suffuses our consciousness and infuses the aesthetic direction of and mentalities behind these games. Due in no small part to its full embrace of capitalist ethos, the Pokémon franchise is one such piece of contemporary pop driftwood. Large and sprawling media enterprises like these are naturally in conversation with global attitudes towards dying. Rather than flee from these adorable Pocket Monsters like the cold light of dawn, we should embrace them in our conversations around death.

Who Rangers Can Not Save

While the games, paratexts, and other media have engaged in discussions of death, the conversation surrounding them has largely lagged behind. Popular perceptions of death in the greater Pokémon universe are largely centred around what could be described as clickbait and listicles. A cursory Google search for 'Pokémon death' returns results like 'Top Ten Anime Deaths' and 'Times Pokemon Actually Died'. The sensationalism here recalls the past discourse around games like *Death Race* – albeit a little more commercially palatable this time around.

Rather than confronting these texts on their own terms, there is an impulse to decry their messages. In 1976, the game *Death Race* was the source of a national gaming scandal. While almost indecipherable by today's standards, in 1976 *Death Race* was a parade of mortal terrors. Dr Gerald Driessen of the National Security Council had this to say 'I shudder to think of what will be next if this one is not defeated by public opinion' (Qtd in Young, 1976). Unbeknownst to Driessen, what would come next would be adorable plush monsters. Though these games and their reactions are divergent, death in the world of Pokémon gets a similarly superficial treatment.

Articles like Baird's (2017) '15 Times Pokémon Actually Died' for ScreenRant and Cooper's (2016) '20 Surprisingly Dark Pokedex Entries from Pokemon Sun and Moon' for Dorkly are entertaining enough, but convey little of the nature of death in the world of Pokémon. Like the veritable *rageposting* and reactionary screeds aimed at videoludic death's more gory representatives, these articles are meant to convey something similarly superficial about death and gaming.

While reactionary commentary on *Death Race* and the history of violent video games after that focusses virtually exclusively on the dangerous and corrupting force of that pixelated mayhem, conversations around Pokémon focus on the spectacle of the game. These encounters with mortality aren't vicious nor are they morally corrosive; they are reduced to bits of intrigue. They are meant to surprise or entertain the reader. The concern here isn't to build a discussion of death within gaming, but to persuade the reader that games are dangerous in their depictions of death. The epistolary onslaught against games like *Death Race* is designed to convince the reader that death in those games poses a real-world threat. The siren song of the listicle likewise calls to the reader, but here is meant to suggest that there is nothing to be seen except for a moment's worth of novelty. This, however, ignores and overly simplifies the interactions with death in the world of Pokémon.

The most infamous example of death in the greater Pokémon franchise-verse isn't in the games themselves, but in the anime. Episode 21 of the *Pokémon Sun and Moon* series titled 'One Journey Ends, Another Begins …' features what has become Pokémon's most serious depiction of death (Pokémon Sun and Moon, 2017).

In the episode, an elder Stoutland (a Pokémon resembling a large Terrier) slowly succumbs to illness and old age. As a viewer, we experience this death through the perspective of the young fire cat Pokémon Litten. The episode is somber and uses the death of a Pokémon to explore the weight of grief and loss. It closes with Litten beginning an adventure with Ash (the main protagonist) having learned Stoutland's Fire Fang attack, suggesting that our lives extend beyond death through the impact we have on the lives of those we care for.

Frank (2017) writes that 'pokémon has rarely broached the subject of death before' and goes on to mention the lighter references to death in the main series games. This echoes Paul Zuniga's (2017) claim that 'there's one aspect of the Pokémon universe which has never been directly addressed: Death'. This, however, is incomplete. While it is certainly true that the Pokémon franchise does not usually give this amount of attention to death and dying, the games and greater franchise are not without approaches to death.

The first episode of the first season of the Pokémon anime, *I Choose You,* reaches its climax with the threat of death. Ash, having just acquired his iconic partner Pikachu, gets in over his head in a battle with a flock of wild Spearow. The *Tiny Bird* Pokémon proves to be more than the pair can handle. Pikachu lays defeated, Ash is wounded and about to sacrifice his life to save his partner, and after a fade out designed to slot in a commercial break, we see Pikachu save the day. The pair awaken the next morning battered and exhausted beyond their limits. The climax of the episode is woven with the threat of death. Like the entire franchise, death is uniformly behind the curtains.

The Fandom.com (2019) wiki for the Pokémon anime franchise has a 'List of Deaths' section. Most of the prominent deaths are 'temporary'. Ash Ketchum, for example, routinely finds his character vaporised, turned to stone, electrocuted, and otherwise dispatched from this mortal coil. Each of these deaths, however, is temporary. His character is always somehow revived – usually with the aid of an intervening and powerful Pokémon. The use of death as a theme is common in Pokémon narratives. It's not just the anime and films that maintain death as a recurring theme, but the games as well.

Pokémon: Magikarp Jump (Select Button, 2017) is the only known instance of you, as the player, being able to directly kill your own Pokémon. In the game, the player exclusively trains a Magikarp via a series of mini games. If the player causes the Magikarp to jump too high, it will be eaten by a hungry Pidgeotto that is flying by. While meta-rulesets such as Nuzlocke runs intimate the death of one's partner Pokémon,[1] *Magikarp Jump* is the only entry in the franchise that allows, or perhaps threatens, the player with the ability to be the direct cause of the Pokémon's death. While other games hint that the player has engaged, or at least is

[1]Nuzlocke Runs are meta rulesets adopted by players to increase both the difficulty and the narrative weight of play. While many variations exist, the two main meta rules are as follows: Players must give each captured Pokémon a nickname – ostensibly to increase their bond with the Pokémon – and if a Pokémon is fainted, it must be released as if it had 'died'.

proximal to, the slaughter of Pokémon by consuming them, as in Slowpoke tails or fried Magikarp, for example, this is the only instance of death being direct.

As alluded to by the '20 Surprisingly Dark Pokedex Entries from Pokemon Sun and Moon' article, the Pokémon game world is suffused with death. Pokédex entries are riddled with contemplations on death. These range from the curiously morbid as in Haunter's entry from *Pokémon Silver Version* (Game Freak, 1999) 'Its tongue is made of gas. If licked, its victim starts shaking constantly until death eventually comes'. There are the contemplative entries as with Primeape's entry from *Sun Version* (Game Freak, 2016a) 'It has been known to become so angry that it dies as a result. Its face looks peaceful in death, however'. There are also outright sinister Pokémon as with Difloon's entry in *Sun Version* 'Stories go that it grabs the hands of small children and drags them away to the afterlife. It dislikes heavy children'. The in-game world of Pokémon may be dotted with PokéRangers and Nurse Joys out to save the day, but these entries nod towards those that cannot be saved.

While these bonus bits of narrative flavour have little-to-no mechanical purpose in the game itself, they nevertheless convey important information about the game world. Contrary to popular perception, Pokémon is not a sterilised game world. While it is decidedly aimed at a younger audience, it's infantilising to suggest the content therein is 'for children'. In that spirit, let's read Pokémon in this same way we read the memento mori of gaming and approach this piece of popular media on its own terms.

Approaching Pokédeath

The main series Pokémon games have found themselves at an interesting cultural moment. The children who grew up playing the first generation of *Pokémon* games for the Game Boy (*Green*, *Blue*, and *Red* versions) are now adults in their own right – many of whom have their own children who are now growing up on later iterations of these very games. Rather than remain penned in by the 'children's game' demarcation, the Pokémon games are now intergenerational and engage an audience ranging from first-time players, hardcore fans, the competitive scene, and everyone in between. With the launch of *Pokémon Go*, the franchise was also able to reach the growing mobile gaming market.

Pokémon Go (Niantic, 2016) is our first stop in approaching death in Pokémon. Not so much for the content of the game, it is an iteration on the traditional gameplay designed to work within the microtransaction heavy mobile gaming arena. Combat has been largely replaced by consumable items, and the game's aesthetics are drawn directly from the rest of the franchise. What makes *Pokémon Go* interesting is that it is the first Pokémon game to have real-world intersections with death and dying.

The first of these crossroads is decidedly real. One of the most salient facts of human mortality is that it is capricious. The *Pokémon Go Death Tracker* (http://pokemongodeathtracker.com/) purports to be a digital record of all the known instances of people dying or being injured while playing the game. The website leads with a tally of the total deaths thus far and lists links to news articles about these injuries. The site is clearly intended as a bit of morbid humour and includes a 'honorable mentions' section tracking absurd, nonlethal *Pokémon Go* news. This includes awarding a '−1' and a '−2' deaths for instances of people saving lives while playing the game. This tracking reminds and underlines that death is integral to the human condition. No doubt similar tallies could be counted for the main series Pokémon games and indeed gaming as a whole. As vehicles for social interaction, games like *Pokémon Go* will have this statistic proximity to real-world deaths.

Pokémon Go makes the social element of Pokémon more explicit than previous games. It is played outside, with groups of people, and walking around populated areas. The Pokémon franchise has, however, always had a social aspect. The first Pokémon games, Red, Blue, and Green Versions (Game Freak, 1996, 1998a, 1998b) started off this social play with Pokémon trading as a built-in feature that required players to connect their games via Nintendo's proprietary 'Link Cable'. The later Pokémon Trading Card Game and Pokémon video game tournaments would only solidify these games as social activities – rather than games played in isolation.

This social element naturally brings these games into dialogue with death. Whether it's *Pokémon Go* having its players be moving around outside or the main series games creating social hubs through forums and other fan-centred sites. One of the most notable additions of *Pokémon Go* is the mapping of Pokémon Gyms and 'Pokéstops' onto real-world locations (Fig. 1). Public works of art, notable buildings, and local landmarks are the most common markers. Another common location that these in-game sites are mapped onto is funeral homes and cemeteries.

Fig. 1. Fan made *Pokémon* Go sign in Helsinki.

The reaction to this has been mixed with some locations calling for the game to be banned on their grounds and others taking a different approach. 'If you own a cemetery location there is a very good chance you have already noticed a large spike in foot traffic on your grounds' writes Ryan Thogmartin (2016). The space between real-world funerary practice and the Pokémon games is not quite as far as it would appear. Writing about MMORPGs specifically, Gibbs et al. (2012) had this to say 'These games provided a social context in life, and they also provide a social context for people's attention to death'. Here, we find *Pokémon GO* serving a similar function by providing social engagement with death.

The article by Ryan Thogmartin for DISRUPT Media goes on to outline the best practices for harnessing the marketing value of *Pokémon Go* for your funeral home or cemetery: 'The first thing to do is find out if your funeral home or cemetery is a *Gym* or *Pokéstop*' (Thogmartin, 2016). The ostensible goal here being to drive traffic and thereby engagement for the more historically minded sites and revenue for others. The article highlights the importance of seizing the opportunity for financial gain that comes with *Pokémon Go*:

> With all the foot traffic Pokemon GO could bring to your funeral home or in your cemetery you should also be considering how to maximise the upside of the extra exposure. (Thogmartin, 2016)

By creating a game with an outdoor, social element that naturally seeks out local landmarks as part of its design, *Pokémon Go* created a new reason for people to engage with death – however incidentally.

Gibbs et al. outlined several forms of funerary memorials in Blizzard's MMORPG (Massively Multiplayer Online Roleplaying Game) *World of Warcraft*. They specifically outline static monuments that resemble real-world gravesites – things like statues and monuments – and later evolutions that added NPC interaction, quest lines, and in-game items that all have ties back to the deceased (Gibbs et al., 2012). While *World of Warcraft* has intentionally made internal funerary monuments and grave sites, *Pokémon Go* has directed people to engage with the same structures in the outside world. The greater Pokémon community's connection to death doesn't stop at incidental tourism.

Posting in the PokéCommunity forums, a user asked the forum questions about Pokémon and death (PokéCommunity, 2010). Forum users speculated as to the nature of death in the Pokémon games. Some suggested that the grasslands were littered with the bodies of defeated Pokémon while others argued that there must be some kind of job in the Pokémon world where people go around saving injured Pokémon. This might seem like a bit of pop ephemera at first. Dedicated fans looking to explain away or dig deeper into the lore. Deeper, perhaps, than one can go. However, if we approach Pokémon on its own terms, a larger discourse emerges.

Writing about the need to incorporate gaming into narrative therapy, Nicolucci (2019, 2) states 'For this generation, video games are one of the most important sources of storytelling'. Pokémon is simultaneously disposable pop culture, children's media, and the stories through which we discuss and frame death. Forum users are concerned with death in the game, because death is an overarching human concern, and this is the media through which it is being explored. Note that similar questions about equally vital 'real world' needs are less prevalent than death. Fewer, or no, posts exist about gastro-intestinal bacteria vital to the digestive process of Pokémon or the chemical makeup of the Pokémon atmosphere. Our inherent concern with mortality is played out through the media in which we participate. Sometimes it is Shakespeare, other times it is Pokémon.

While the forum users at PokéCommunity present the possibility for in-world healing for fainted Pokémon, whether by other *healer* Pokémon or as a job for humans in the Pokémon universe, other forum users had a much more stark outlook. Their replies paint the opposite picture and argue that the phraseology 'Faint' and 'Fainted' were just Nintendo's way of saying 'dead'. (PokéCommunity, 2010). Other comments speculate about Pokémon eating wild Pokémon that have been injured in battle[2] or the existence of a Poké-afterlife where Ghost Type Pokémon are created (PokéCommunity, 2010). With varying amounts of 'lore' to support their arguments, these forum users are attempting to work out and negotiate an approach to death from within a culturally shared art form. Pokémon serving here are both a hobby of sorts and as a shared story through which people are forming understandings of death.

Users of the GameFAQs (2011) forums openly pondered on a more naturalistic approach to death in the world of Pokémon when they asked if other wild Pokémon consumed the bodies of those the trainer defeated. In these two forum posts alone, everything from theistic metaphysics to strict materialism has been expressed. Posting in the same *Pokémon Black Version* (Game Freak, 2010) forum, a user argued that while Pokémon likely die, it wasn't something the player actively did as part of the game (GameFAQs,2011). We even see the proximal unease we have with death played out in the videoludic. An 'unreal' space is mapping directly on to 'real' anxieties.

Despite, or perhaps because of, its ongoing status as a 'children's' game, the Pokemon franchise has always engaged in frank, earnest discussions about Death. The exploitative nature of 'mature' discussion slipped past these games. Rather than euphemistically trite platitudes 'Pikachu has gone to live on a farm upstate' or the frankly insulting 'Don't worry, Cubone's Mother is just asleep for a very long time'. These games approach the realities of death with a direct approach.

Returning to Zuniga's (2017) review of the Pokémon anime. We find an interesting comment: 'What's interesting about Stoutland's death is that there was a lot of heavy emotional weigh in the scene.' The Pokémon universe is clearly not without appraisals of death and the culture we build around it. As we have noted, the prevalent attitude that Pokémon somehow skips out on death is misleading. Further, the notion that emotionally waited deaths are somehow new to the games is equally misleading.

---

[2]There is a fair amount of in-game support for Pokémon eating other Pokémon. The Pokédex entry for Toxapex reads: 'Toxapex crawls along the ocean floor on its 12 legs. It leaves a trail of Corsola bits scattered in its wake' (Game Freak, 2016a).

As far back as the original games, released in 1996 in Japan and 1998 in the United States, heavily weighted deaths have been key thematic elements for the games. The original games feature a monument known as Pokémon Tower. Situated in the corner of the eerie Lavender Town, the Pokémon Tower is home to the Pokémon franchise's first cemetery. The tower is the site of gravestones, ghosts, and the possessed. Iconography aside, Pokémon Tower is most notably home to the story of Cubone's Mother.

The character of Cubone's Mother has grown over the iterations with 2018's *Pokémon: Let's Go, Pikachu and Let's Go, Eevee* being the latest updates (Game Freak, 2018a, 2018b). The lore tells the story of a mother Marowak defending her young child Cubone from an attack by the game's recurring villains Team Rocket. While this death is never displayed on screen, Marowak gives her life to defend her child.

If the player attempts to climb to the top of Pokémon Tower before acquiring a key item, the Silph Scope, later in the game, they will be stopped by an unbeatable ghost that directs them to 'Get out …' In the earlier games, the player would have to battle and defeat the Marowak, proving that they are strong enough to kick Team Rocket out of the town. In the more recent renditions, the games have merged with the story as presented in the anime. Upon using the Silph Scope to reveal the ghost as the deceased Marowak, the orphaned Cubone appears and embraces the ghost of its mother. Able to find peace knowing her child is safe, the Marowak departs for, presumably, the afterlife and the player is able to defeat Team Rocket and bring the town to peace.

The player doesn't drive away an unclean spirit in the manner of an exorcist, but assists the dead – and their surviving loved ones – in healing through restorative justice. The player actively participates in this grieving process. The game cannot continue on from this point unless the needs of the dead are met. Rather than simply frame the undead Marowak as a violent ghost, easy enough for video games to accomplish, the most recent version seeks instead to present our contact with dead loved ones as something that can bring healing. This presents the deceased not as inanimate or hostile bodies, but as agential and in continued possession of their best qualities from life. The encounter with Cubone's mother is an emotional highpoint for the game that feels as if it could be at home in games purpose built to engage with death.

Philippa Warr's (2017) review for PC Gamer brings to light an important distinction: *A Mortician's Tale* is centrally about death, while Pokémon is centrally about adventuring with, battling, and collecting fantasy monsters. Pokémon, however, brings death into its larger preview without being a game 'about death'. Our discourse around death and dying in the videoludic need not be limited by the express intention of a game's developers. Games, such as *A Mortician's Tale, That Dragon, Cancer,* or *What Remains of Edith Finch,* are masterful explorations of the end of life through the medium of games, but we lose something in the discourse when we focus so tightly on these games. Culture and media are inexorably intertwined and as long as Pokémon remains one of the most widespread and successful video games, it will continue to be a medium through which people look to understand death.

The Pokémon franchise is sprawling, and there is room in that growing media empire for the serious and somber appraisal of death. The Pokémon games fit into the history of the videoludic exploration of death as seen in both popular titles and independent games. The fandom and culture accreting around Pokémon is likewise engaged and is seeking not only to understand how these games approach death, but how the medium and texts therein can help them to understand death and dying more broadly. With each new iteration, these games add to this mortal discourse and invite us to find new ways to encounter death and dying in the Pokémon universe.

Life in the Shadow of Pokémon Tower

We live in the shadow of Pokémon Tower. Just as Lavender Town rests in the shade of this looming memento mori, so does all media exist in a context framed by death. As we have seen throughout all Pokémon media, death is ever present. Sometimes it is playful, sometimes horrific, often somber, but always there. As Caitlin Doughty (2016, 113) wrote in *Smoke Gets in Your Eyes and Other Lessons from the Crematory:* 'Death might appear to destroy the meaning in our lives, but in fact it is the very source of our creativity .... Death is the engine that keeps us running'. This embrace of death suffuses and threads through all media – no matter how *pop* they might appear.

Despite often being depicted as a binary state, death remains experientially fluid in gaming. The Pokédex entry for Yamask from *Pokémon Black* (Game Freak, 2010) marks this like a semantic headstone: 'Each of them carries a mask that used to be its face when it was human. Sometimes they look at it and cry'. Pokémon, like Yamask itself, is a mask that allows us to confront and interact with death. Sometimes we look upon the mask and cry, other times we wear the mask and play games. Pokémon invites us to contemplate the inexorable.

As we are compelled by the text to 'Catch 'Em All' so too are we driven to connect death, play, and Pokémon.

Interactions and Sensoria

Exidy. 1976. *Death Race*. Exidy.

FromSoftware. 2011. *Dark Souls*, Namco Bandai Games and FromSoftware.

ID Software. 1992. *Wolfenstein 3D*, Apogee Software and FormGen.

Oda, Jason. 2014. *Continue?0987654321*.

White Owls Inc. 2018. *The Missing: J.J. Macfield and the Island of Memories*, Arc System Works.

Yager Development. 2012. *Spec Ops: The Line*, 2K Games.

References

Anable, A. 2018. *Playing with Feelings: Video Games and Affect*, Minneapolis, MN, University of Minnesota Press.

Baird, S. 2017. 15 times pokémon actually died. *ScreenRant*, 15 June. Available at: http://bit.ly/2Ww0hxI

Cooper, T. 2016. 20 surprisingly dark pokedex entries from pokemon sun and moon. *Dorkly*, 16 August. Available at: http://bit.ly/31WElwB

Debord, G. 2005. *Society of the Spectacle*. The Anarchist Library.

Doughty, C. 2016. Happy 5th Anniversary, Deathlings! *Order of the Good Death*, 28 January. Available at: http://bit.ly/2WpErvJ

Durkin, K. F. 2003. Death, dying, and the dead in popular culture. In *Handbook of Death and Dying*, Ed. C.D. Bryant, pp. 43–49. Vol. 1, Thousand Oaks, CA, Sage Publications, Inc.

Fandom.com. 2019. List of deaths Wiki. *Pokémon Wiki*, 15 May. Available at: http://bit.ly/36lBpNk

Fishman, J. M. 2017. *Death Makes the News: How the Media Censor and Display the Dead*, New York, NY, New York University Press.

Frank, A. 2017. Pokémon sun and moon anime reminds us that Pokémon definitely die. *Polygon*, 7 April. Available at: http://bit.ly/36nnhDE

FromSoftware. 2014. *Dark Souls 2*. Bandai Namco Games.

Funeral Direct. 2017. Ever dream about being a funeral director? This game lets you play as one. *Funeral Direct*, 15 December. Available at: http://bit.ly/2NvmAPI

GameFAQs. 2011. What happens to fainted wild Pokémon? *GameFAQs Forum*, 15 March. Available at: http://bit.ly/2Nq2NBd

Game Freak. 1996. *Pokémon Green*, Nintendo.

Game Freak. 1998a. *Pokémon Red*, Nintendo.

Game Freak. 1998b. *Pokémon Blue*, Nintendo.

Game Freak. 1999. *Pokémon Silver*, Nintendo.

Game Freak. 2010. *Pokémon Black*, The Pokémon Company and Nintendo.

Game Freak. 2016a. *Pokémon Sun*, The Pokémon Company and Nintendo.

Game Freak. 2016b. *Pokémon Moon*, The Pokémon Company and Nintendo.

Game Freak. 2018a. *Pokémon: Let's go, Pikachu!*, The Pokémon Company and Nintendo.

Game Freak. 2018b. *Pokémon Let's go, Eevee!*, The Pokémon Company and Nintendo.

Gibbs, M., Mori, J., Arnold, M., and Kohn, T. 2012. Tombstones, uncanny monuments and epic quests: memorials in world of Warcraft. *Game Studies*, *12*(1).

ID Software. 1993. *DOOM*. List.

Laundry Bear Games. 2017. *A Mortician's Tale*, Laundry Bear Games.

Muncy, J. 2017. A Mortician's tale might be the first game to really get death. *Wired*, 20 October. Available at: http://bit.ly/2JAaEuO

Niantic. 2016. *Pokémon: GO*. Niantic.

Nicolucci, V. 2019. A death-positive video game for death education of adolescents. *Italian Journal of Educational Technology*, *27*(2), 186–197. https://doi.org/10.17471/2499-4324/1071

Numinous Games. 2016. *That Dragon*, Cancer. Numinous Games.

PokéCommunity. 2010. What happens to fainted Pokémon? *PokéCommunity Forums*, 26 August. Available at: http://bit.ly/33b0dpH

Pokémon Sun and Moon. 2017. One journey ends, another begins.... SM021,6 April, Asada, Yuji.

Russell, S. 1962. *Spacewar*! Steve Russell.

Select Button. 2017. *Pokémon: Magikarp Jump*. The Pokémon Company.

Thogmartin, R. 2016. Embracing Pokemon go: A how to guide for funeral homes and cemeteries. *Disrupt Media*, 13 July. Available at: http://bit.ly/2BVaN7U

Warr, P. 2017. A Mortician's tale review. *PC Gamer*, 18 October. Available at: http://bit.ly/36kUWO0

Young, L. 1976. Local safety authorities denounce game. *The Spokesman-Review*, 29 December. Available at: http://bit.ly/2Pt51CG

Zuniga, P. 2017. Latest Pokémon Anime depicts first canonical death—so what happens after they die? *Futurism*. Available at: http://bit.ly/2Nre6sX

Chapter 3

Staying Dead: The Corpse, Burial and Exhumation in Three Contemporary British History Plays

Benjamin Poore

Abstract

This chapter examines the acts of burial and exhumation in three contemporary British history plays. For the purposes of this argument, a 'history play' may be defined as a piece of writing for the theatre that engages with historical events or settings. Such plays inevitably, at the moment of their staging or revival, take on particular meanings for audiences, since theatre as a live, durational art form encourages spectators to compare the historical events depicted with their present historical moment. The chapter argues that acts of burial and exhumation in contemporary British theatre are intimately tied to notions of land, soil and belonging. These became increasingly pertinent ideas in the UK's political climate in the years following the 2016 Referendum on membership of the European Union. Of the three case studies, *Victoria* by David Greig (2000) dates from more than a decade before this vote, whilst *Common* by D. C. Moore (2017), and *Eyam* by Matt Hartley (2018) were written and staged in the interim between the Referendum result and the UK's withdrawal from the EU. All three, however, feature corpses on stage as a means to consider time, temporality, place and history. Each play offers a different interpretation of what it means to play dead and to stay dead.

Keywords: Theatre; drama; playwriting; corpse; folk horror; soil

Introduction

This chapter examines the acts of burial and exhumation in three contemporary British history plays. For my purposes, a 'history play' may be defined as a piece of writing for the theatre that engages with historical events or settings. Such plays

Death, Culture & Leisure: Playing Dead, 41–53

doi:10.1108/978-1-83909-037-020201010

inevitably, at the moment of their staging or revival, take on particular meanings for audiences, since theatre as a live, durational art from encourages spectators to compare the historical events depicted with their present historical moment. The chapter will argue that acts of burial and exhumation in contemporary British theatre are intimately tied to notions of land, soil and belonging. These became increasingly pertinent ideas in the UK's political climate in the three years following the 2016 Referendum on membership of the European Union. Of the three case studies, *Victoria* by David Greig (2000) dates from more than a decade before this vote, whilst *Common* by D. C. Moore (2017) and *Eyam* by Matt Hartley (2018) were written and staged in the interim between the Referendum result and the UK's withdrawal from the EU. All three, however, feature corpses on stage as a means to consider time, temporality, place and history. Each play offers a different interpretation of what it means to play dead and to stay dead.

Recent work in theatre studies has highlighted the numerous and extensive connections between death and the theatre. The language of death is 'part of the performance vernacular', according to Curtin (2019, 5) in his book *Death in Modern Theatre*: we speak of 'corpsing' and 'knocking 'em dead', of 'kill[ing] it' and mounting a 'revival'. Robson (2019) notes the superstitions around death among actors and discusses instances of actual death and near-death onstage. There are some indications that the earliest kinds of theatre and performance were rituals connected with death (Robson, 2019, 10), and certainly one of the earliest plays in the classical canon, Sophocles' *Antigone* (c. 441 BC) is intimately concerned with death and burial rites (Robson, 2019, 58–60). This ongoing connection may be because 'theatre allows us to memorialise the dead and make them feel present to us [...] and by extension, to contemplate death and ponder our mortality' (Curtin, 2019, 12) and also because

> Theatre exposes the cultures of death to contemplation, to a thoughtful looking [...] The culture of death is another name for 'living-with' others, so long as we recognise that those we live with include not only those present but also those no longer living and the not yet living. (Robson, 2019, 67–68)

Death in the theatre can help us to situate ourselves in culture, among the living and among the dead, and in time. As Geraldine Cousin (2013, xi) remarks:

> Theatre is ideally suited to the representation of precariousness because the theatrical, present moment is itself inherently unstable [...] The passage of time is built in to the occasion, and it is precisely for that reason that time itself can become part of the playwright's game plan.

As this chapter will show, the invoking of death through the figure of the corpse almost inevitably leads to a distorted temporality in the play text.

At the same time, a dead body on stage also involves a risk that it will be read as comic rather than profound. Robson (2019, 23) discusses the dead body's 'capacity of a combination of horror, abjection and laughter' via Joe Orton's

Loot, first produced in 1965, and featuring the embalmed body of Mrs McLeavy, which is removed from its coffin and hidden in a cupboard so that the coffin can be filled with loot from a bank robbery. Curtin (2019, 7–9) cites John Dryden's 1668 comments warning against actors dying onstage since 'the Audience cannot forbear laughing', and also notes the physical awkwardness of actors having to play dead convincingly, particularly for long stretches. As I will show, this comic potential is exploited in Matt Hartley's *Eyam* when the body of Unwin appears to come back to life while being buried (Hartley, 2018, 130).

Hence, the comic is just one of the alternative modes that can be accessed, accidentally or on purpose, by the appearance of a corpse in a play performance. However, the modes that this chapter will focus on in detail are Mark Fisher's theories of the weird and the eerie, and Adam Scovell's theorisation of folk horror. In a previous publication, my co-authors and I attempted to align one of the plays featured here with the contemporary Gothic (Jones, Poore and Dean, 2018, 7–11). This chapter proposes an alternative theorisation that seeks to account for the distorting theatrical effects of the corpse and the grave. Each of the plays under detailed consideration in this chapter has one or more historic and rural settings, which is in keeping with folk horror's strong association with the rural (Scovell, 2017, 9) and the past: 'the past and the present mix and create horror through both anachronisms and uncomfortable tautologies between eras' (Scovell, 2017, 10). Similarly, for Fisher's (2016, 11) definition of the eerie, isolated locations are significant: '[a] sense of the eerie seldom clings to enclosed and inhabited domestic spaces: we find the eerie more readily in landscapes partially emptied of the human'. For Scovell (2017, 184), the EU Referendum activated a disturbing kind of folk horror in everyday life: 'there have been few political events in recent years that have resembled the genre in such startling detail'. Referencing the ritual murder of the police officer in the seminal folk horror film *The Wicker Man* (Hardy, 1973), and comparing it to what he sees as the irrationality and inwardness of pro-Brexit views, he adds: "We have burnt our Sgt. Howie in the wicker man, and now wait naively for our apples to grow once more, confident that we have 'taken back control'" (Scovell, 2017, 184). Particularly in the case of *Common*, I will argue, these scripts play on an ambivalence about land and soil in a national sense: is the land presided over by largely benevolent *genii locorum*, or guardian spirits, or is it desacralised, haunted by devils (Legard, 2018, 381), a refuge or a threat?

The Object of Possession

At the root of much folk horror seems to be a fear of the land itself:

> The danger can be from those who work the soil or those who want to kill for the soil, from things excavated from under the soil, or indeed from the very soil itself. (Scovell, 2017, 38)

Consequently, many folk horror narratives involve the unearthing of a powerful object – 'the object of possession' – that contains some kind of evil; this trope, in turn, has its roots in the macabre fiction of the fin de siècle (end of the century)

(Scovell, 2017, 42 and 31). As an example of how this trope features in contemporary playwriting, we might consider *Sanctuary* by Tanika Gupta, first performed at the National Theatre, London, in 2002. Set in 'a small Eden-like' garden in the corner of an urban graveyard (Gupta, 2002, 15), the play is about the parish priest, Jenny Catchpole, and the local inhabitants who use the garden as a place of safety and peace. At the play's climax, the church gardener Kabir discovers that Michael, who had fled to the UK from Rwanda five years earlier, was actually a perpetrator of atrocities during the Rwandan genocide, rather than a victim as he had always claimed. This knowledge comes to light when Kabir unearths a cigar box, buried by Michael, containing documents revealing his true identity. As a consequence of this discovery, Kabir hacks off Michael's arm with a machete and sets fire to the shed with Michael in it. When the other characters discover the burned shed containing a charred body the next morning, Jenny and her grandmother lead the digging of a shallow grave in which to place Michael's remains before the police arrive, in order to protect Kabir. The play's title, therefore, refers to the garden itself as a sanctuary, but also makes reference to the places of sanctuary and refuge sought by victims of the Rwandan genocide, including churches and schools, in which they were executed, as recalled in the play's second act (Gupta, 2002, 93). In Rwanda, sanctuary is denied to the victims, while in the UK it is denied to the perpetrator.

In *Sanctuary*, then, a past is symbolically buried in a graveyard – perpetrator Charles Bagilishema changes his identity to victim Michael Ruzindana – and safety is sought in an urban Eden. Sebastian, an alcoholic journalist, has been trying to find Charles Bagilishema to confront him with his crimes for eight years, and it is he who asserts Michael/Charles' guilt before Kabir unearths the evidence. As a result, the violence of Rwanda is visited on Michael in an English graveyard and his body burned, then buried, in a reversal of what is done to Oscar in *Victoria* (and in anticipation of the immolation of the caravan at the end of Jez Butterworth's play, *Jerusalem*).

The trope of the cursed object in the soil is also reversed, in a different way, in Mike Bartlett's state-of-the-nation allegory, *Albion*, first staged at the Almeida Theatre in London in 2017, where a historic garden in Oxfordshire is being restored to its post-World War I glory by the play's protagonist, Audrey. She has scattered her son's ashes in the garden, and for her birthday is given a trowel by her husband, Paul, that had been owned by the architect of the garden, Weatherbury (Bartlett, 2017, 90). Here, the cursed object is the digging implement rather than what is dug up, since Audrey comes to the realisation that there is nothing but decay in the soil: 'Groundwater. Disease in half the trees. Miner bees everywhere. Ants. There's some things you can't restore. The earth itself' (Bartlett, 2017, 113). Perhaps surprisingly, then, *Sanctuary* seems more optimistic in its view of English soil: it gave up Michael's secrets and then swallowed him in return. The native soil of Albion, designed by a survivor of the Great War, is not nourished by the ashes of Audrey's son James, a soldier killed by an improvised explosive device in Iraq. James' partner tells Audrey bitterly, 'And then you scattered him in this mess. In this indulgent, waste of time. You ruined his life. Now you've ruined his death too' (Bartlett, 2017, 119).

Each of the three plays that are discussed in detail in this chapter was produced by a high-profile institution that might be seen as nationally representative, or as representing the UK to the rest of the world: the Royal Shakespeare Company (RSC), the National Theatre and Shakespeare's Globe. Like *Albion*, two of the three – *Common* and *Eyam* – were programmed at a time when a historical play would very much be expected to have said something about the state of the nation now; the earliest play, *Victoria*, was staged at the turn of the millennium, again a point at which a major, subsidised national company might be expected to reflect, through its programming choices, on the nature of history and progress.[1]

Victoria

David Greig's *Victoria*, first performed by the Royal Shakespeare Company at the Barbican in London in 2000, is an example of a play that uses both rootedness to place and distorted temporality. Set in 'a rural place on the coast of Scotland' and in three historical periods (1936, 1974 and 1996), the play features three characters named Victoria in each of these different generations; the play script stipulates that '[t]he character of Victoria must be played by the same actress throughout' (Greig, 2000). The first Victoria emigrates, pregnant and having been let down by her lover, Oscar, in 1936. The second Victoria ('Vicky') is the sole survivor of a helicopter that crashes into the mountainside in 1974. The third Victoria is the wealthy daughter of Vicky and Euan, a ruthless businessman. In the first part of the play, Oscar – Euan's father – shoots David, the Nazi-fellow-traveller son of the local landowner, Lord Allan, and makes his way to Spain to join the International Brigades. Despite an apparent lack of closeness in life (Greig, 2000, 137 and 147), Victoria is haunted by Oscar's ghost (171–172). Indeed, earlier, at the funeral, Euan had been confronted by his father's ghost:

Euan *is at the grave.*
Oscar *is sitting on it.*
Euan Vicky thinks this place is haunted.
Thinks there's memory in the stones.
Thinks she sees fucking events, historical events.
I own this place.
I own the fucking sea and the fucking mountain and the forest.
This place is mine.
So don't ... don't ... come back.
I own these fucking historical events which take place in Vicky's fucking head in this fucking place.
So don't come back.
Close my eyes and I don't want to see you.
Close my eyes and you'll be gone.
Oscar *doesn't move.* (149–150)

---

[1]See also my article on the RSCs new work season of the following year, 2001, and the way that these plays also reflected on the national past and present (Poore, 2018).

Victoria, who is withdrawing from her psychology degree at Glasgow University and looking for direction in life, persuades her friend Billy to help her disinter Oscar and burn his body so that his ashes can be scattered in Spain, as he had wished. This tension between belonging simply to the soil of one place and the urge to get away – even to leave one's heart elsewhere – is captured by Victoria's lines to her grandfather's ghost:

> Stone too heavy to carry.
> Can't go forward.
> Me and all that made me.
> The weight of oldness.
> This place.
> If I could step out of my body.
> Leave my skin and bone on the hillside.
> Walk into some new life.
> But I'm half in and half out of the ground.
> Place holds me.
> The shape of me.
> In the mountainside. (172)

What makes this Victoria's ritual all the more complex is that she initially wants to have sex with Billy and to give him a large sum of cash – 'all the money I could take out the bank' (178). Billy rejects both of her offers, and so the money is burnt on top of Oscar's body, along with his memoir of the Spanish Civil War. The act of burning a large amount of cash might be taken as a reference to The K Foundation (artists Bill Drummond and Jimmy Cauty), who in 1994 famously burned a million pounds on the Scottish island of Jura (Brook 1997, 5). In the play, the deliberate destruction of both money and of a valuable historical record can be read as Victoria's (and even the play's) rejection of a false 'heritage' narrative that seeks to recuperate the revolutionary spirit of people like Oscar into capitalism. In one scene, a publicist, Kirsty, is shown speaking at a council meeting and offering, on behalf of Euan, to build an Oscar Sutherland Business Training Centre in return for planning permission for a second granite quarry (a literal hollowing out of the mountainside). Euan, Kirsty reports,

> feels such an initiative will do justice to the memory of his father, whose life was dedicated to increasing opportunity for all, and to the community in which he himself was brought up. (171)

This is clearly meant to be read as a cynical distortion of Oscar's socialist principles to serve his son's commercial agenda.

In the play's final moments, the first Victoria returns from Argentina and voices her own sense of being 'Like a stone half in the ground' (180). Nevertheless, it is hard to read the end of the play in a purely positive light; Oscar's Civil War memoir records moments of horror and humiliation as much as inspiration, and we are shown how even Oscar appropriated the testimony of his comrade

Euan (164). Victoria, meanwhile, seems to have her own reasons for disinterring and burning her grandfather's body, and they are neither based on political nor family loyalties; earlier she had said to Billy: 'Nobody should have to do anything. That's what I believe. What's that? To believe that? Is it communist?' (168). Watching the body burn, she says to Billy: 'This is what I look like. This is what the world looks like [...] After the fire' (181). Her wish to go to Spain with Billy to scatter Oscar's ashes reads more like an escape or a moment of self-discovery than a deeply-felt reverence for historical truth or Oscar's principles.

Temporality

Plays like *Victoria* that are concerned with the persistence of place are in some sense obliged to draw attention to their temporality in order to do so. Nick Stafford's *Luminosity* and Moira Buffini's *Loveplay* were both staged by the RSC at the Barbican the following year to *Victoria*, and both were set on a patch of land, where events occur decades and centuries apart (Poore, 2018). Simon Longman's *Gundog*, produced by the Royal Court Theatre, London, in 2018, and, set in an unidentified rural landscape amid much talk of blood and soil, reveals its unusual treatment of time midway through the first scene, where sisters Becky and Anna are questioning a man found wandering the fields, Guy Tree:

> **Anna** Reckon he'll be here this time next year?
> **Becky** Have to see.
> *A year passes.*
> **Becky** Guess he really doesn't have anywhere else to be.
> (Longman, 2018, 20)

Mike Bartlett's *Albion*, mentioned earlier, accelerates its temporality at the end – having previously hinted at temporal estrangement through fancy dress and the appearance of a ghost – so that in the final moments 'the ground is returned to soil' and the 'house is destroyed' (Bartlett, 123). In Rory Mullarkey's *Saint George and the Dragon* (Mullarkey, 2017), 'home' grows from a rural village – the production's stage design literalising the idea of a 'green and pleasant land' and 'the sunlit uplands' – to a thriving Victorian town to a dystopian contemporary city, over three acts. The same characters live on across the centuries. Richard Bean's *Harvest* (Royal Court, 2005) presents the history of a family farm in East Yorkshire, won from the local Lord in a bet, across different decades of the century, and more, that followed.

Common

Victoria features a character who is buried and then removed from his grave. In *Common*, however, one of the play's central events is the protagonist, Mary, digging *herself* out of her grave. Mary returns to her home village at the beginning of the play, having been presumed dead for many years and having made her fortune in London. She seeks to reclaim her lover Laura from their brother, King, who had tried to kill Mary, an act that led to her leaving the village in the first place.

From the beginning, *Common* plays on the tropes of vengeful, torch-bearing villagers, who are seen in the Prologue making 'Rough Music' (Moore, 2017, 5) as they cut open and rip out the heart a of pig that is dressed as the local Lord. The play is set in 1809, at the time of the Enclosures, and prior to the pig slaughter the villagers had been ripping out fence-posts that marked the edge of the common. Soon after this, Mary remarks to the crow-scarer Eggy Tom that it

> was once Custom in this Land to build a vast straw wicker man
> fill its frame with all the young boys and girls from a village then
> burn them alive incased inside to please Taranis, God of Thunder.
> (Moore, 2017, 10)

This speech, of course, references in supercharged form one of the foundational films of folk horror, *The Wicker Man*. As the quotation suggests, *Common* is written in a distinctive language with the capitalisation of initial letters reflecting eighteenth and early nineteenth-century typographic practices, and with run-together words and missing punctuation providing clues for emphasis in performance. The language implies a historical past, but not a realistic version of one, more a dreamed past in which the dead insistently return.[2]

In addition to slaughter and sacrifice, the idea of burial is introduced at an early point. We first meet Laura as she drags a slaughtered foal onstage, which she is burying to propitiate the earth after a poor harvest (Moore, 2017, 25). When Laura and Mary meet and Mary reveals her identity in the dark, she says that saying her own name is 'like hearing spell cast from deep belly of the Earth' (36). It transpires that King, Laura's stepbrother, tried to kill Mary some years ago, but she escaped the village and King told everyone that she was dead (36–37). Mary and Laura meet on the common at night as Laura is unearthing Mary's coffin, which contains one of Mary's dresses, and hay, dirt and rags to give the impression that it contained a body (67). But by the end of the scene, instead of them arranging to emigrate to America as Mary had proposed, as the villagers gather round Laura stabs Mary and King pushes her into the open grave, a double betrayal by both brother and lover.

Mary's self-exhumation at the start of Act Three is anything but low-key. She is 'bloodied and muddied', and an orange glow rises from the coffin hole. She then has a conversation with a talking crow named Eggy Robert, who tells her that 'Crows see all through time & place [...] what lies behind & futurefront' (Moore, 2017, 77). She then swears that she will institute 'our next Moments of Pure Reckoning'; she is pregnant and means to provide for her child by any means necessary (Moore, 2017, 79). She raises her hands and is struck by lightning. The next

[2]Both *Eyam* and *Common* make use of crows as sinister figures associated with death; where *Common* in the National Theatre production used projection and animatronics, *Eyam* insists that the crows be played by humans (Hartley, 2018, 15).

scene has the feel of a fairy tale, with Mary waking up in the Lord's 'second-best bedroom' and attended by her maidservant Young Hannah, who (the stage direction tells us) 'most resembles Eggy Tom' who had died earlier (Moore, 2017, 80).

Hence, where for *Victoria* death is complicated by the existence of ghosts, in *Common* the characters Mary and Eggy Tom appear to be unkillable, reappearing as soon as they die in either the same or an altered form. *Victoria*, written at the turn of the twenty first century, is preoccupied with the legacy of the struggle against fascism seventy years earlier. *Common* responds to the 2016 EU Referendum by emphasising the hidden savagery of 'darkest England'.[3] And yet it is hard to fit *Common* into any cohesive political position. Mary might at first appear to be an avenger figure, either representative of the disenfranchised working class or of feminist revolt against a patriarchal rural society. But Mary's embrace of cruelty – for instance, as she beats the victimised Irish labourer Graham to the ground, seemingly without motive (103) – and her siding with Lord against the villagers' protests, suggest a calculating immorality. She also teaches Young Hannah to behave with cruelty to servants as they prepare to go to London together (113). She betrays those who trust her, as she in turn has been betrayed, including Lord, who asks (as Graham had earlier), whether she is the devil (102 and 116). Indeed, in the language of folk horror, she might herself be read as a cursed object of possession, one who digs herself up.

In Fisher's terms, it is a play that is eerie in its evocation of landscapes that are being emptied out, empty coffins and empty graves, and yet it is also a series of displacements and substitutions that fill the empty spaces: a pig replaces Lord, Connor the Irish labourer replaces the pig, Mary's body replaces her empty coffin, Young Hannah replaces Eggy Tom and Mary kills and buries her murderer, King. In the closing moments of the play, Mary is represented as a *Zeitgeist* figure as she heads for London:

> I will build, burn, make money dance to My Will in that Finecity. And from its river-arteries reach out into the Growingworld & take what is Mine, as the World stole my Blood from Me. (116)

As a parting shot, she tells the audience that the world they are in is 'That which I'll make next' (118). If there is an interpretation that we are encouraged to make here, it seems to be that of the displaced working class seeking new lands through empire, stealing in return for what was stolen from them. Hence, if it is to be read as an allegory of nation, of land and soil, it is a profoundly ambivalent one, even more so than *Victoria*.

Eyam

Matt Hartley's *Eyam*, which premiered at Shakespeare's Globe Theatre in September 2018, is my final example of a play that explores history through death

[3]I use this phrase in reference to my argument on the post-imperial Gothic in Jones, Poore, and Dean (2018).

and burial, and this time the play in a very particular landscape rather than the unnamed rural Scotland and rural England of *Victoria* and *Common*. The village of Eyam in Derbyshire became famous for the inhabitants' decision to quarantine themselves during an outbreak of bubonic plague in 1665–1666. Hartley himself grew up within sight of the village ('Eyam', Hartley, 2018) and therefore felt a duty to the landscape and its inhabitants, living and dead, in constructing the play. The published play text is preceded by four pages of character biographies for the historical figures represented, and Hartley has stuck, presumably for reasons of historical veracity, to the names on the historical record, an anomaly acknowledged in the script by William's exasperated line,

> Elizabeth! This village appears to be populated solely with Johns or Elizabeths or Marys. Is it a trick that everyone is playing upon us and really all go by different names? (Hartley, 2018, 40)

This joke directed towards the audience or play reader also points to an underlying tension in the narrative. The new reverend at Eyam, William Mompesson, and his wife Katherine discover that their predecessor, Shoreland Adams, was lynched and then dragged through the lanes tied to a horse. One of the villagers, John Hancock, points out to William and Katherine: 'Eyam is a village cut off from the world. It's an island. Where isolation lives: discord, rivalry, friction, they all breed. It can become a law unto itself' (Hartley, 2018, 35). This warning also, of course, indicates an affinity with the folk horror trope of strange earth rituals and the mysterious vanishing or murder of outsiders, as also featured in *Common*. Unlike *Common*, however, *Eyam* was a Globe production for its open-air amphitheatre, and made no use of eerie lighting effects. The dominant costume colour was black, and the Globe stage was boarded over in black. Like *Saint George and the Dragon* mentioned earlier, the village setting was realised by a group of miniature model houses. The lip of the stage was extended by two v-shaped catwalks, facing in towards each other and creating a central triangle – as well as numerous other angles – for groundlings to stand in. Despite the non-naturalistic production design, graves were indicated by patches of soil on the stage surface.

Even before the arrival of the plague, Eyam is shown to be a village at home with corpses and burials (indeed, Hartley may be indicating that people in the seventeenth century had this much more open attitude towards death). The first person that the Mompessons meet when they arrive at the beginning of the play is the sexton, and a local girl, described by others as a 'ghoul' who spends too much time in the graveyard, plays with a dead woman's finger (Hartley, 2018, 21). Discussion then turns to the dissenting minister, Reverend Stanley, who is legally barred from returning to the village but whose wife wished to be buried there (Hartley, 2018, 32). As with *Common* and *Victoria*, land in the play is threatened by the capitalist appropriation and privatisation of natural resources. Philip Sheldon, the wealthiest person in the village, is seeking to buy up everyone's land so that he can control the mining of lead in Eyam (Hartley, 2018, 37). As Mary Talbot says – in a line that could equally have been said by characters in *Common* and *Victoria*, 'To leave my land is to rob my own children of their history' (Hartley, 2018, 37).

The plague arrives via a consignment of damp cloth sent to a tailor, Viccars, who is lodging with Mary Cooper and her son Edward. The play implies an attraction between Edward and Viccars, so that the notion of freedom – leaving home, embracing a gay sexuality – is mixed with the idea of death. 'There is a door. Follow me through', Viccars' last words, take on a different meaning when he 'unbuttons his shirt' and 'He is covered in boils' (Hartley, 2018, 77). Edward's unwillingness to give up Viccars' body – sitting with the corpse for days and telling his mother that Viccars had left – provides the occasion for the villagers to fight over it, whereupon it is dropped, the sheets open and the audience witnesses a rotten body, covered in flies, and with an overpowering smell (that feature of the unburied body that Laura Wade's play *Breathing Corpses* also dwells on) (Hartley, 2018, 80).

The other individual burial and corpse that is given particular attention in *Eyam* is that of the poacher Unwin, who becomes an entrepreneur – a sort of disaster capitalist – when the plague spreads. Like Euan in *Victoria*, he preaches the proto-Thatcherite gospel of hard work and identifying new business opportunities. He styles himself as 'a man who creates work [...] This is the time to endeavour, to build, to cut waste' (108). When it is Unwin's turn to be buried, the sexton Howe makes clear to Mary Cooper that Unwin will not be mourned or missed, but as he shovels the earth onto the grave, Unwin begins to stir. He is kicked and beaten by Howe, either because he believes that the body is possessed or in order to check that Unwin is alive (130–131). Restored to life, Unwin subsequently refuses to continue his quack-remedy business with Harriet; he is a changed man, born again (134–135). Thus, the Thatcherite entrepreneur is allowed to appear in Eyam, but Hartley's investment in the essential goodness of this place does not permit him to succeed. Instead, Sheldon stands in as both the industrialist and the pre-capitalist robber baron figure; he leaves Eyam as soon as the plague is discovered and is evidently not 'of the people': the character notes tell us unequivocally that he is 'a bastard who revels in being a complete arsehole' (Hartley, 2018, 9).

Unwin's return from the dead is treated as comedy, slapstick even, but it is bookended by a more sombre story thread, that of Mary Talbot digging a grave for a second of her children, and then finally for herself as she too catches the plague. Her act is presented as one of some moral complexity: Mompesson has instructed everyone in the village to stay indoors and not to assist with neighbours' burials. As John Hancock says to Mompesson: '[You] Ask us to deny our most basic of instincts' (Hartley, 2018, 128). Mompesson, in turn, offers to help dig her grave himself. When Mary is on the verge of death, she seems to speak to the Globe audience:

> My name is Mary Talbot. This is the land our heavenly father did bestow upon me. My husband I buried here. My children I bury here. I will now rest here. And I will rest well, for this land I take to my grave.

The speech refers explicitly to a sense of belonging in the soil, of her death after those of her family having a satisfying closure to it. It is quite unlike the folk horror conventions of wicked deeds buried in the soil, objects of possession and the blood-sacrifice of outsiders. Here, the bodies are returned willingly to their own land.

A final indication that Hartley's play seeks to reverse the tropes of folk horror rather than to simply invoke them is that the play ends with William reciting the statistics of deaths in the village, and insisting, 'They must not be written from history./ They must not be forgotten./ They will be remembered' (140). William then recites the names of everyone who died, until 'The entire village is with him. A community' (143). This act of recital fits squarely with Robson's comments on how mass deaths call for new ways of thinking and staging collective experience (Robson, 2019, 42). Speaking the victims' names can be seen here as a way of avoiding the 'sense of the deaths of others as anonymous deaths which cannot and need not be grieved, mourned, or atoned for' (Robson, 2019, 42). The sense of this marking, de-anonymising and commemorating the dead at the end of the performance was enhanced, on the day I saw it, by the tossing of roses onstage in the final moments. Though I did not realise it at the time, this is a Globe tradition for the final performance of any show (Cantoni, 2017, 107), but in the moment, it read as an additional invitation to think of the Globe stage as a transported Eyam, and the flowers as tributes on the victims' graves.

Conclusion

There are many other important contemporary British plays that feature burial and exhumation, but this chapter has aimed to compare the use of these tropes in contemporary political drama, and across three contemporary history plays in particular. In *Victoria*, Oscar 'plays dead' in a way that is conventional for theatre – he appears as a ghost – although his being laid to rest is an ambiguous gesture by the third Victoria. In *Common*, Mary plays dead but refuses to stay dead; she rewrites the villagers' sacrificial ritual for her own ends, consigning King and Laura to the grave. In *Eyam*, Unwin represents the comic possibilities of playing dead: both comic in the sense of slapstick, and in the generic sense of his being reintegrated into the community as a result of his experience. Read as commentaries on the modern nation in 2000, 2017 and 2018, these plays are contrasting treatments of rural Scotland and England as 'home', as 'heartland', as 'native soil'. The 'half in and half out of the ground' of *Victoria* is rendered literal onstage by *Common*, where at various points King, Laura and Mary are all standing, alive, in open graves (77, 91, and 92). Yet *Victoria*'s sense of complex and partial belonging through the soil is made more straightforward in Mary Talbot's declaration of 'this land' in *Eyam*. The three historical periods of *Victoria* point to a loss of ideals, a rise in individualism (literally) eroding the landscape and a modern sense of being cut adrift. *Common*'s dreamlike vision of history evokes Scovell's nightmarish feeling about Brexit and folk horror; the play replicates some of Brexit's tensions between the local and the global, between the country and the city, but chooses not to resolve them. Lastly, *Eyam* offers a narrative of death uniting a community, resolving tensions between 'native' and 'outsider', in a way that implies a similar, if forlorn, hope for a united kingdom today.

Interactions and Sensoria

Butterworth, J. 2009. *Jerusalem*, London, Nick Hern.
Doran, G. (Director). 2009. *The Royal Shakespeare Company Production of Hamlet.*
Gimpo (Director). 1995. *Watch the KLF Burn A Million Quid.* Available at: https://vimeo.com/21103501 [Accessed on 1 September 2019].
Hardy, R. (Director). 1973. *The Wicker Man.*
Kane, S. 1996. *Blasted & Phaedra's Love*, London: Methuen Drama.
McGrath, J. 1981. *The Cheviot, the Stag and the Black, Black Oil*, London, Methuen Drama.
Stafford, N. 2001. *Luminosity*, London, Faber and Faber (print).
The Fall. 1988. "Jerusalem." *I Am Kurious Oranj*. By W. Blake, H. Parry, M.E. Smith, J. Cohen, Beggars Banquet.
A. Paciorek, G. Malkin, R. Hing and K. Peach. 2018. *Folk Horror Revival: Field Studies*, 2nd ed., Durham, Wyrd Harvest Press.
Wade, L. 2005. *Breathing Corpses*, London, Oberon.

References

Bartlett, M. 2017. *Albion*, London, Nick Hern.
Brook, C. 1997. *K Foundation Burn a Million Quid*, London, Ellipsis.
Cantoni, V. 2017. *New Playwriting at Shakespeare's Globe*, London, Bloomsbury Methuen Drama.
Cousin, G. 2013. *Playing for Time: Stories of Lost Children, Ghosts and the Endangered Present in Contemporary Theatre*, Manchester, Manchester University Press.
Curtin, A. 2019. *Death in Modern Theatre: Stages of Mortality*, Manchester, Manchester University Press.
Fisher, M. 2016. *The Weird and the Eerie*, London, Repeater.
Greig, D. 2000. *Victoria*, London, Methuen.
Gupta, T. 2002. *Sanctuary*, London, Oberon.
Hartley, M. 2018. *Eyam*, London, Nick Hern.
Jones, K., Poore, B. and Dean, R. 2018. Introduction. In *Contemporary Gothic Drama: Attraction, Consummation and Consumption on the Modern British Stage*, pp. 1–20, London, Palgrave Macmillan.
Legard, P. 2018. The Haunted Fields of England: Diabolical Landscapes and the Genii Locorum. In *Folk Horror Revival: Field Studies*, Eds A. Paciorek, G. Malkin, R. Hing and K. Peach, pp. 378–389, Durham, Wyrd Harvest Press.
Longman, S. 2018. *Gundog*, London, Bloomsbury Methuen Drama.
Moore, D.C. 2017. *Common*, London, Bloomsbury Methuen Drama.
Mullarkey, R. 2017. *Saint George and the Dragon*, London, Bloomsbury Methuen Drama.
Paciorek, A., Malkin, G., Hing, R., and Peach, K. 2018. *Folk Horror Revival: Field Studies*, 2nd ed., Durham, Wyrd Harvest Press.
Poore, B. 2018. Before the Fall: Looking Back on the Royal Shakespeare Company's "This Other Eden" Season (2001), *Journal of Contemporary Drama in English*, 6(1), 179–190.
Robson, M. 2019. *Theatre & Death*, London, Red Globe Press.
Scovell, A. 2017. *Folk Horror: Hours Dreadful and Things Strange*, Leighton Buzzard, Auteur.

Chapter 4

Death, Playfulness and Picture Books

Maggie Jackson

Abstract

This chapter offers a brief overview of the ways in which death has been addressed in children's picture books in a playful or light-hearted manner. The books here are a small purposive sample showing the 'ordinary' way, in which death can be dealt with in picture books rather than looking at books with a specific therapeutic intent. The concepts of 'playfulness' and also 'carnival' are explored before four books are analysed.

Keywords: Picture books; playfulness; death; children; carnival; ordinary

Introduction

> dealing with death has remained problematic
> in the UK and 'Western' societies more generally.
> (Hester and Taylor, 2011, 200)

Addressing the subject of death with children is still often regarded as a contentious task. Moss (1972, 530) commented 'we ply children with pamphlets, films and gorgeously illustrated books on birth, while we ignore death'. Earlier Anthony (1940, x) commented that 'we find an obstinate refusal to accept the fact of natural death'. This notion that death is not part of the 'normal' life experience (for children, in particular), allows talk about death to be seen as a matter for specialist intervention and one best avoided if at all possible. Indeed, Gutiérrez et al. (2014, 43) found in their research that the common belief held by most adults is that children should be shielded from death. Haynes and Murris (2013) also noted this tendency to restrict or censor what children can access, assuming they will be unable to understand, or be distressed by the notion of death. This despite evidence that 'death is very much a part of the fantasy thought of children' (Bluebond-Langer, 1977, 171); others would agree and suggest it should be

Death, Culture & Leisure: Playing Dead, 55–66

doi:10.1108/978-1-83909-037-020201011

part of the cultural lives of all, including children (Ribbens, 2006). Picture books can be an accessible way of introducing difficult subjects with young children and offer a 'versatile medium capable of conveying a narrative at a variety of levels' (Graham, 2005, 221). Many children's books being written since the 1980s and 1990s addressed 'adult' concerns and have not shied away from serious subjects and thus can be seen as a useful source through which to address 'challenging themes' (Lampert and Walsh, 2010, 148).

A Brief Overview of Books on Death for Children

Gibson and Zaidman (1991, 232) noted however that 'when books expressly for children first appeared in English, they were often about death'. Since the 1980s, there has been an explosion in the publication of children's books addressing the subject of death. Often, though, these books fall into the category of 'bibliotherapy' and serve to help children cope with bereavement and thus tend to concern themselves with the message rather than enjoyment. Often where these books address death they do so in a simple but direct manner, even for the very young, using concepts of universality, irreversibility, non-sentience and inevitability (Speece and Brent, 1992); these being deemed to be key to a mature understanding of death though books for younger children tend to concentrate on non-sentience and irreversibility. The clear message being that death is something different from life – it is not like sleep and you cannot recover from it. This might seem like a harsh message to present to children, but directness and clarity are helpful to allow the young child to process complex and difficult ideas. Singer and Singer (1992, 179) go as far as to suggest that the effect of reading stories to children can be one of 'reducing the world to a manageable size'.

Dick Bruna (2005) in his small book *Dear Grandma Bunny* tells his very young readers on page three that grandma Bunny was not just sleeping: 'she is much stiller than before' and also that 'she isn't breathing any more'; in fact, she is dead he tells us. Although this book could not be said to be 'playful' as such, the character Miffy is well known to the young and might be said to part of the everyday life of many children.

Books dealing with difficult subjects can sometimes focus on the message rather than on being just a good story and thus can risk lacking in appeal to the young. This is echoed in the suggestion that the aesthetic and literary merits of books which deal with 'difficult' subjects are sometimes sacrificed because of the nature of the subject in question (Lampert and Walsh, 2010). Though death had never been entirely absent from children's literature (Jackson, 2018) in some ways, it has begun again to become a subject 'fit' for inclusion in the literature for the young and very young (Moore and Mae, 1987). Some (Flood, 2018) lament this and would suggest that as a result books have lost the aspect of joy and wonder of earlier years.

However, in the nineteenth century Lear's limericks often addressed the subject of death with humour (Butler, 1972), and in a similar way, in the early twentieth century Belloc (1907) continued to use humour and death in his *Cautionary Tales for Children*. Belloc's characters did not all die as a result of their misdemeanours, but many did meet with an untimely end, notably *Jim* 'who ran away from his nurse and was eaten by a lion' (Belloc, 1907, 149). From the first decade of the twentieth century death (in an ordinary way) began to be viewed as a less suitable subject for children's books – at this time, they were written for a 'recognizable childhood … and any didactic intent … was a poor second to entertainment' (Hunt, 1994, 59). This peaceful world of the late Victorian and early Edwardian eras seen in the books written at the time contained what has been described as 'useful fictions' (Myers, 1989, 52) which served to reinforce the apparently new view of childhood as a time of play and innocence. It is important to note that the prevailing societal attitude towards the notion of 'childhood' will be the one reflected within the books produced, and thus deemed suitable, for children (Jones, 2006). Thus in seeking to shield children from the realities of the world 'we fill their environments with primary colors and fun and cuddles, and purge those environments of any references to despair or struggle' (Barthlemess, 2013, 62).

However, many of the books written since the 1980s and 1990s have addressed 'adult concerns' (Postman, 1994) such as death and loss and as a result have not shied away from serious subjects, despite the notion that certain subjects, often featuring animals or toys, are usually deemed more suitable (Nikolajeva, 1998). It would be fair to say that many of these books aimed at children from the age of three upwards could not be said to be humorous exactly; many take a playful approach to death and thus allow difficult subjects to be considered in an engaging and non-threatening way. 'Sad books are more difficult to find in the picture book canon, amidst the boisterous alphabets and sunny barnyards and irreverent pigeons' (Barthlemess, 2013, 56), but playful picture books addressing those difficult subjects do exist and can help children in processing the difficult emotions and concepts.

Here, I consider four examples of children's picture books, published between 1987 and 2011, and explore the ways in which they introduce ideas of death with light-heartedness and humour. The four were selected from a larger sample from previous research (Jackson, 2018). The books chosen here illustrate a view one might describe as 'continuing bonds' theory (Klass et al., 1996); they each include happy memories of the deceased and thus allow 'humour' to be used. They mostly show that death is final and irreversible, a concept that is important in addressing the subject of death with children (Speece and Brent, 1992). As others have also found animals feature quite often in books for younger children (Jones, 2001). Thus, this selection serves to illustrate how death can be addressed with young children (from four years up) in playful, light hearted and ordinary ways.

The notion of playfulness will be explored and applied to each of the books in order to analyse the ways, in which this helps to make a 'difficult' subject accessible, ordinary and, perhaps, even fun. The books considered in the remainder of this chapter include:

- *Lola and the Rent-a-Cat* (Jitta, 2007);
- *Bone Dog* (Rohman, 2011);
- *Dog Heaven* (Rylant, 1995); and
- *Fred* (Simmonds, 1987).

The use of pictures alongside words can allow for a certain level of agency to be afforded to the child 'reader', allowing them to explore the subject on their own terms and to be curious and questioning about what they see, whether they can understand the written word or not. As such, picture books allow for meaning making through the combination of image and text. The use of the image as a significant part of the narrative and not merely as an illustration of the words can offer a broader and deeper story – we should note that in the picture book 'pictures are never just pictures' (Lewis, 2001, 74). The combination of words and pictures leaves a space in which the child can explore and make sense of even the most difficult of subjects and where this can be done with humour (or a playful approach) and can provide a safe space for the child to explore. We might also consider that it allows the adult some relief from the anxiety of the subject being seen as too 'heavy'. Rather they can make the subject ordinary and so 'books that speak to sadness – that acknowledge its very existence – help establish loss and grief as part of the human condition' (Barthlemess, 2013, 56).

Although death is not a subject that one might readily associate with the notion of playfulness and humour, particularly in relation to picture books for young children, it has been noted by Serafini and Coles (2015, 636) that 'the use of humor helps decrease anxiety'. Others would suggest that it is in the picture book that serious subjects are now being addressed with humour. Picture books do not have to rely only on words alone but the 'illustrations can provide material which is not explicit in the text' (Pinsent, 2007, 169) and thus can open up a subject to the child without the need to be able to read words. This in turn means that they can be engaged with without adult intervention. The use of the image as a significant part of the narrative and not merely as an illustration of the words can offer a broader and deeper story. Lewis (2001, xiii) considers this to be a way of telling a story which is shared 'between two different forms of communication', but which as a result offer more. Others (Sipe, 2012, 7) suggest that the pictures make the story 'come alive', whereas Iser (1978) considers that there is a space left by the text that allows the pictures to offer new understandings.

Playfulness

The concept of playfulness can be somewhat hard to define (Lieberman, 1977) and particularly so when it is not being observed as a behaviour but through the secondary agency of a book. Nonetheless, a working definition of what it might be will be required here. In combination with the notion of playfulness, I will also borrow from the ideas of Bakhtin and his use of the idea of 'carnival' in order to explore ideas of subversion and the ways in which 'taboo' or apparently 'tasteless' subjects have been addressed within the picture books analysed here. Using humour does not make a subject less serious but 'is just as admissible in … literature, posing universal problems, as seriousness' (Bakhtin, 1984, 66).

Lieberman (1977, 127) considers that playfulness is the 'rearrangement of the familiar', a theme echoed in Bakhtin, and what Dewey (1910, 162) describes as 'an attitude of mind'. The concepts of playfulness and carnival each contain the notion of role reversal, in that the natural order can be subverted; the fool can be king, the child wiser than the adult and this allows the apparently forbidden areas of life to be considered, and even mastered, by those usually excluded from such subjects: thus children and death.

As a way of addressing 'difficult' subjects humour – or playfulness – can be a useful and gentle way of allowing children to think about and know about death as an 'ordinary' subject that can be seen and read about within the everyday books available to children.

A playful attitude towards addressing subjects can be an 'ideal blend' (Skilbeck, 2017, 5), as it allows for enquiry into new worlds and also for the possibility of exploring the unknown or difficult in a safe way. Further he suggests that it is a way of questioning 'assumptions about a topic' (Skilbeck, 2017, 5). James (2004, 377) also noted how death has often been addressed in children's books with humour and thus allows for it to be 'transformed and conquered'. It is this notion of allowing the child to play with a subject and thus gain mastery (Lieberman, 1977) that we can see within the picture books to be considered here. Three of the books are concerned with the death of a pet and often this may be a child's first experience of death (Formanek, 1974). Each of these shows how death can be dealt with as an 'ordinary' subject (that is part of life) as part of an engaging narrative, whilst also addressing sadness and loss in a light-hearted and entertaining way.

I will now go on to explore how death and playfulness has been addressed in the four books chosen and will also bring out themes used by others (Green, 2008; Poling and Hupp, 2008; Corr, 2004; Jones, 2001) who have also researched children's books to show the illustrate the 'educational' function of these books. Jones (2001) used four broad themes which I have used in a modified form elsewhere (Jackson, 2018) and apply them here to some extent. The themes are: the announcement of death, reactions to the announcement, funerals/rituals and the afterlife and memory/continuing bonds. The themes serve merely as a frame to refer back to the 'educational' function of the books. I will consider each of the books and show how these themes apply in terms of death content and also consider and compare the ways, in which playfulness has been used within each theme.

Four Examples of Playfulness in Picture Books

Fred offers a very direct account of the death of Fred. On the first page (note that picture books are rarely numbered), the children (Nick and Sophie) with whom Fred lives are seated in front of their door. They meet one of the neighbours and say 'Our cat, Fred, has died'. Immediately the death is announced; we are in no doubt about what has happened. Then there are three more frames, in which the children tell those they meet the reason for Fred's death – he was old and had cat flu. A full page picture of the deceased (whilst still alive) follows before moving on to scenes of the burial – the page showing how Fred was in life can be read almost as a eulogy – ' he liked eating – but most of all he liked sleeping – on the washing – on top of the fridge'. Clearly, Fred was a much loved cat – and though it is going too far to say these statements are funny – they share the same wry tone often heard at funerals. The next two pages are then given over to the actual burial in the garden – incidentally near the graves of two other pets. The children discuss what to write on Fred's grave stone – the little boy suggests saying 'he was the laziest cat' though his sister vetoes this and they simply write Fred – as she explains – 'you have to write something nice'. The children in *Fred* (Simmonds, 1987) are very worldly and wise; they speak in an adult way about Fred's death – the juxtaposition of their very small bodies looking up at the adults to whom they tell their story draws on the Bakhtinian notion of 'carnivalesque'. Here, roles are reversed, the fool is wise, the king is low and children can be more grown up and knowledgeable than the adults. Of itself this may not seem funny in terms of belly laughs but the mimicking of 'adult' behaviour and conversational style allows for role reversal and thus humour.

Lola and the Rent-a-Cat (Jitta, 2007) is not about the death of a pet but of Lola's husband. However, this does reflect the tendency in children's books to more commonly address either the death of a pet or an older person (Ordal, 1980). The first four pages give us a brief overview of the lives of Lola and her husband together and then on page five we have one line opposite a full page drawing which states 'one day John falls over, his heart stops beating' – these are the only words on the page opposite the picture of John upside down having fallen off his chair. As with *Fred* we are left in no doubt about John's fate. The words and pictures are clear and to the point. The 'carnival' aspect in this story comes also from the subverting of the expected. Lola is an 'old woman' with grey hair in a bun but the pictures of the life she and her husband lead prior to his death seek to challenge our assumptions of old age. John and Lola are shown in the bath together – Lola scrubbing John's back (though there is no real reason to assume this is not usual in real life); they walk hand in hand, they swing on the backs of their chairs whilst reading the newspaper and – Lola wears training shoes. Subversion of the stereotypically expected causes us to smile at least, if not laugh out loud. John's death is graphic – but humorous in its directness and slightly shocking crudeness – he is shown having keeled over on his chair whilst Lola peers down at him still stuck to the chair. Bodily functions including the crudest depictions of death are part of the carnival spirit – they allow for the fear (of death) to be conquered and brought low. What we see is a dead body – but in an undignified position (Jitta, 2007, 6).

Bone Dog (Rohman, 2011) is in some ways less direct, not using the word 'dead' but rather the dog (Ella) tells her owner Gus 'I'm old and won't be around much longer' and then next we are told that Ella is 'gone'. In some ways, this echoes the teaching of the *Ars Moriendi* which contained instruction on the right way to die and was available to be read by the literate laity (Binski, 1996, 39). It was a 'How to' guide for the ordinary sinner. Knowing how to die well was an important factor in the life of the medieval person, including children (Kastenbaum, 2004): Ella shows here that she knows how to die well. This is also within the first pages of the story. Unlike the previous two books, no humour is offered to temper the sadness of the loss at this stage. Gus trudges about his daily life with hunched shoulders and a general unwillingness and lack of desire to be happy. He has to go to a Halloween party which he doesn't want to go to 'but he pulled on his costume and trudged out of the door'. The humour and subversion in this story comes later and in the form of 'magical' thinking, or escapist thought.

The final book, *Dog Heaven* (Rylant, 1995), does not mention that the dogs are dead but starts by telling us that 'when dogs goes to heaven, they don't need wings because God knows that dogs love running best' – we see a little dog walking up a hill on one side to the empty space of the hill on the opposite page. In a sense, the death is announced by the fact of being in heaven. Although this book only looks at the afterlife, it does so in a subversive, carnival-like way. The pictures are crude drawings, as if done by a child but they have movement – each running over two pages and allowing the child reader to scan for unexpected images.

Lessons in Loss and Death

I will now go on to consider how loss and death have been addressed in each of the books. Only one of the books – *Fred* (Symonds, 1987) – considers the funeral explicitly and indeed the whole book is given over to the wild and raucous funeral service the cats give for Fred. This contrasts with the more sombre affair we see in the early pages where the children, with their parents, bury Fred in the back garden. This is a homely and reassuring image, but once the parents are removed from the narrative life becomes subversive and much more exciting. Cats from all over begin to arrive; the children are woken during the night by a cat in traditional funeral attire asking if they are 'friends of the dear departed'. They then discover that not only was Fred their cat but that he was known to others as 'Famous Fred' and that cats have come from far and wide to mourn and to celebrate his passing. As can happen at a funeral, we find out things about someone that were previously unknown to us. Here, the children discover that Fred was a famous singer and the cats tell stories about his life, then sing and cry and throw flowers on the grave, then go on to the funeral 'tea'. The sheer exuberance of the pictures makes this event one of joy; though solemn and must allow for children to ask about funerals in general. The contrast of the occasion with the cats and mice singing echoes this notion of picture books being able to 'poke fun at everyday events and illuminate the humorous aspects of the human condition' (Serafini and Coles, 2015, 636). It is plain to see here that funerals are not always sad affairs.

Both *Bone Dog* (Rohman, 2011) and *Dog Heaven* (Rylant, 1995) are concerned with notions of afterlife and continuing bonds (Klass et al., 1996). Poling and Hupp (2008) found in their research that children's books tend to refer to some form of afterlife, although as with the two books here the conflate the notions of afterlife and remembering, perhaps reflecting more contemporary ideas of the place of the dead in what we might term 'Western society' (Hester and Taylor, 2011; Ribbens, 2006; Walter, 1999).

The dogs in *Dog Heaven* (Rylant, 1995) have fun being dead. Heaven is full of all their favourite things; there are geese to bark at, children to throw balls for them and God bakes biscuits for them shaped as squirrels and cats. This playful idea of heaven is shown by the simplicity of the illustrations which are all two pages; thus allowing for movement across the page as we see the dogs running freely in the wide open spaces of heaven. The memories or continuing bonds

(Klass et al., 1996) described in this book are not those of the owners but of the dogs. The concept of the continuing bond (Klass et al., 1996) here is reversed by telling us that the dogs remember their owners on earth and that 'sometimes an angel will walk a dog back to Earth for a little visit and quietly, invisibly, the dog will sniff about his backyard' (Rylant, 1995, 16). The owners can't see the dog, but the dog is reassured and 'when he is satisfied that all is well, the dog will return to Heaven with the angel' (Rylant, 1995, 20). The image shows a happy dog being led by an angel bounding up a brightly coloured stairway back to heaven. Here, we smile at the pictures, we do not laugh out loud but even so notions of 'carnival' and the disruption of the normal order allow for difficult ideas and emotions to be addressed.

Bone Dog (Rohman, 2011) is perhaps slightly more painful and wistful in tone. We sense the real loss Gus feels for Ella – on the second page Ella says 'no matter what happens I will always be with you'. Clearly, this is not a promise that can normally be kept; thus this book allows for wishful or 'magical' thinking when Ella returns to rescue Gus from some menacing skeletons and is the least helpful in terms of understanding death as it ignores the concept of irreversibility (Speece and Brent, 1992). Gus is out on his own on Halloween, dressed as a skeleton, when he meets a host of skeletons who dance about him and want him to join in their fun. When he reveals that he is a real boy and not a skeleton they turn on him and try to kill him – 'you've got guts kid … but not for long' (Rohman, 2011, 12). There is humour here; there is also danger and subversion. Ella returns to rescue Gus – she is smaller than the menacing skeletons who merely laugh at her – but when she and Gus howl they call up an army of real dogs who chase the skeletons away. The boy and the bone dog outwit the skeletons and stay united as friends forever. Gus asks is he will see Ella again – perhaps suggesting some awareness that death is final and irreversible – but the reply is rather more magical and reassuring.

Lola and the Rent-a-Cat (Jitta, 2007) uses the notion of subversion, as we saw in *Fred* (Symonds, 1987). Playing on stereotypical expectations about an older woman the notions of death and loneliness are explored and addressed by offering unexpected ideas about how she might react to the death of her husband of 56 years. At first, Lola just goes through the motions of doing the everyday things she always did (on pages 8 and 9), but we can see that they no longer satisfy her. The words tell us 'The days are long. There is no one to look after anymore' (page 9). The words are sad but in amongst them we are told she surfs the internet – perhaps not so very odd but challenging our expectations about an old lady with grey hair in a bun. The familiar tropes about old ladies and cats are used here, and then subverted in the 'carnivalesque' manner when we see Lola surfing the net for a suitable cat – like a dating agency. Pages 12–13 show the profiles of a range of cats she had viewed. Having chosen 'Tim' he arrives the next day and Lola's life begins anew – but the final pages take us back to the continuing bond (Klass et al., 1996) with her husband. Tim helps her but has not replaced he beloved husband – 'Lola stokes Tim. She thinks about the old days' (page 19). The following pages show pictures of all the things she remembers form her life. The humour, such as it is here, is the juxtaposition of the sadness of loss, the incongruity of an 'old lady' with her wrinkled skin and grey hair, her trainers, surfing the net and also the challenge to remember she was once young and in love.

Concluding Remarks

What this small sample has shown is that it is not uncommon, though perhaps for some a little surprising, to note the directness of address in these picture books for young children (Jones, 2001). Poling and Hupp (2008) noted in their research picture books aimed at younger children tended to focus on what they term, 'biological aspects' of death (i.e. the fact of death), and this may reflected here in the use of direct reference to death rather than the use of euphemism. We know straightaway, within the first pages, what has happened. Picture books, unlike those with just words, function in a performative manner. We are not required merely to imagine the characters and settings but are presented a visual representation, as in a play (Sipe, 2012). Often when picture books are read to small children, the reader will also 'perform' the characters, using multiple voices to recreate the characters. This then allows for playfulness to be an essential feature of the picture book experience and thus 'engages the interest of the student, encourages inquiry … unsettles prior knowledge and opens up students to the possibility of new knowledge' (Skilbeck, 2017, 5).

When considering how this is manifested in the picture books chosen here, it can be seen that taking the young reader beyond their expected knowledge is key part to the stories. As we see in *Lola and the Rent-a-Cat*, the 'truth' about the pain of loss is made clear but we can still have fond memories; this is done in a light-hearted manner and allows us to at least smile. This book is less boisterous than 'Fred' and though the humour is gentle nonetheless the light-hearted tone allows for 'everyday fears' (Serafini and Coles, 2015, 637) to be addressed and thus taken seriously. This would seem to echo Haynes and Murris (2013, 1084) who suggest 'we cannot seriously engage with children in authentic searches for understanding if we have already decided that children lack true authority'. That is to say children can make sense of the world if we allow them to know about it.

McDaniel (2001) noted that the literary merit of books which deal with 'difficult' subjects sometimes make sacrifices in terms of aesthetic and literary merit, however these stories are engaging and humorous and not merely didactic – though they do not shy away from being factual and direct. The use of humour to make death 'manageable' is not new (Stevanović, 2009), as mentioned earlier, nor is it ever entirely absent from children's books (Jackson, 2018). The books here do not offer laugh out loud images around death and dying; that would perhaps be inappropriate, particularly for the young audience for whom these books are intended. They do show a playfulness and way of facilitating enquiry without fear of the subject being too serious to question.

These books do not avoid death, not even the fact of universality (Speece and Brent, 1992; Jackson, 2013). Even in a book as apparently simple as *Dog Heaven*, pages 4 and 5 tell us that 'there are children, of course in heaven' and illustrate this with images of angel children on bikes chasing dogs up and down hills, playing with kites and throwing balls for the dogs – and generally having a fun time; everyone dies – even children. Death here is colourful and not static.

What we have seen in this small sample of picture books is that even where the text is somewhat dull and plain, the pictures are 'exuberant and humorous' (Sipe, 2012, 9) and thus allow for the exploration of 'everyday fears that children experience' (Serafini and Coles, 2015, 637). Perhaps unsurprisingly animal deaths feature in three of the four books considered here (Moore and Mae, 1987; Ordal, 1980). Done with humour and a playful manner, it is thus possible to 'look at death within the ordinariness of everyday experience' (Coombs, 2014, 2).

Interactions and Sensoria

Burningham, J. 1990. *Granpa*, London, Red Fox.
Gorey, E. 1998. *The Gashlycrumb Tinies*, London, Bloomsbury Publishing.
Hay, S. 2013. *Undead Pets: Night of the Howling Hound,* London, Little Tiger Press.
White, E.B. 1963. *Charlotte's Web*, London, Puffin Books.

References

Anthony, H.S. 1940. *The Child's Discovery of Death*, London, Kegan Paul, Trench, Trubner and Co.

Bakhtin, M.M. 1984. *Rabelais and His World*, Translated by H. Iswolsky, Bloomington, IN, Indiana University Press.

Barthlemess, T. 2013. What makes a good picture book. *The Horn Book Magazine*. 89(5), 56–62.

Belloc, H. 1907. *Cautionary Tales for Children*, London, Eveleigh Nash.

Binski, P. 1996. *Medieval Death*, London, The British Museum Press.

Bluebond-Langer, M. 1977. *The Private Worlds of Dying Children*, Princeton, NJ, Princeton University Press.

Bruna, D. 2005. *Dear Grandma Bunny*, London, Egmont.

Butler, F. 1972. Death in children's literature, *Children's Literature*, 1, 104–124. https://doi.org/10.1353/chl.0.0649

Coombs, S. 2014. Death wears a T-shirt listening to young people talk about death, *Mortality*, 19(3), 284–302. https://doi.org/10.1080/13576275.2014.916257

Corr, C. 2004. Introduction to death-related literature for children: a special issue for omega, *Journal of Death and Dying. OMEGA*, 48(4), 291–292. https://doi.org/10.2190%2FREKE-D7R9-667G-DM79

Dewey, J. 1910. *How We Think*, Boston, MA, Health and Co.

Flood, A. 2018. Domestic dramas are crowding out adventures, warn children's book prize judges, *The Guardian*, 3 May. Available at: http://bit.ly/2BL9atF

Formanek, R. 1974. When children ask about death, *The Elementary School Journal*, 75(2), 92–97.

Gibson, L.R. and Laura, M.Z. 1991. Death in children's literature: taboo or not taboo?, *Children's Literature Association Quarterly*, 16(4), 232–234. https://doi.org/10.1353/chq.0.0855

Graham, J. 2005. Reading contemporary picture books. In *Modern Children's Literature: An Introduction*, Ed. K. Reynolds, pp. 209–226, Basingstoke, Palgrave MacMillan.

Green, J.W. 2008. *Beyond the Good Death: The Anthropology of Modern Dying*, Philadelphia, PA, University of Pennsylvania Press.

Gutiérrez, I. T, Miller, P.J., Rosengren, K.S. and Schein, S.S. 2014. Affective dimensions of death: Children's books, questions, and understandings, *Monographs of the Society for Research in Child Development*, 79(1), 43–61. https://doi.org/10.1111/mono.12078

Haynes, J. and Murris, K. 2013. The realm of meaning: imagination, narrative and playfulness in philosophical exploration with young children, *Early Child Development and Care*, 183(8), 1084–1100. https://doi.org/10.1080/03004430.2013.792256

Hester, R. and Taylor, W. 2011. Responding to bereavement, grief and loss: charting the troubled relationship between research and practice in youth offending services, *Mortality*, 16(3), 191–203. https://doi.org/10.1080/13576275.2011.586121

Iser, W. 1978. *The Act of Reading: A Theory of Aesthetic Response*, Baltimore, MD, Johns Hopkins University Press.

Jackson, M. 2013. Children, grief and the understanding of death. In S. Kreitler and H. Shanun-Klein (Eds.), *Studies in Grief and Bereavement*, Hauppauge, NY, Nova Publishers.

Jackson, M. 2018. *An Exploration of Death in Children's Literature 1890–2010*. Ph.D. thesis, Durham, Durham University.

James, K. 2004. Subversion or socialization? Humour and carnival in Morris Gleitzman's texts, *Children's Literature in Education*, 35(4), 367–379.

Jitta, C.J. 2007. *Lola and the Rent-a-Cat*, Belgium, Baeckens Books.

Jones, E.H. 2001. *Bibliotherapy for Bereaved Children: Healing reading*, London, Jessica Kingsley Publishers.

Jones, K. 2006. Getting rid of children's literature, *The Lion and the Unicorn*, 30(3), 287–315. http://doi.org/10.1353/uni.2006.0033

Kastenbaum, R. 2004. *On Our Way: The Final Passage through Life and Death*, Berkeley, CA, University of California Press.

Klass, D., Silverman, P.R. and Nickman, S.L. (Eds.). 1996. *Continuing Bonds: New Understandings of Grief*, New York, NY, Routledge.

Lampert, J. and Walsh, K. 2010. 'Keep talking until someone listens': understanding prevention concepts in children's picture books dealing with child sexual abuse, *Children's Literature in Education*, 41(2), 146–167.

Lewis, D. 2001. *Reading Contemporary Picture Books: Picturing Text*, Oxford, Routledge Falmer.

Lieberman, J.N. 1977. *Playfulness: Its Relationship to Imagination and Creativity*, New York, NY, Academic Press, Inc.

McDaniel, C. 2001. Children's literature as prevention of child sexual abuse, *Children's Literature in Education*, *32*(3), 203–224. https://doi.org/10.1023/A:1010402202633

Moore, T.E. and Mae, R. 1987. Who dies and who cries: death and bereavement in children's literature, *Journal of Communication*, 37(4), 53–64. https://doi.org/10.1111/j.1460-2466.1987.tb01008.x

Moss, J.P. 1972. Death in children's literature, *Elementary English*, 49(4), 530–532.

Nikolajeva, M. 1998. Exit children's literature?, *The Lion and The Unicorn*, 22(2), 221–236. https://doi.org/10.1353/uni.1998.0028

Ordal, C.C. 1980. Death as seen in books for young children, *Death Education*, 4(3), 223–236. https://doi.org/10.2190%2FW04T-7CAN-LT3B-XBQD

Pinsent, P. 2007. *Children's Literature and the Politics of Equality*, Abingdon, David Fulton Publishers.

Poling, D.A. and Hupp, J.M. 2008. Death sentences: a content analysis of children's death literature, *The Journal of Genetic Psychology*, 169(2), 165–176. https://doi.org/10.3200/GNTP.169.2.165-176

Postman, N. 1994. *The Disappearance of Childhood*, New York, NY, First Vintage Books.

Ribbens, M.J. 2006. *Young People's Experiences of Loss and Bereavement: Towards an Interdisciplinary Approach*, Maidenhead, Open University Press.

Rohman, E. 2011. *Bone Dog*, New York, NY, Roaring Brook Press.
Rylant, C. 1995. *Dog Heaven*, New York, NY, The Blue Sky Press.
Serafini, F. and Coles, R. 2015. Humor in children's picture books, *The Reading Teacher*, 68(8), 636–638. https://doi.org/10.1002/trtr.1361
Simmonds, P. 1987. *Fred*, London, Andresen Press Ltd.
Singer, D.G. and Singer, J.L. 1992. *The House of Make-believe*, Cambridge, MA, Harvard University Press.
Sipe, L.R. 2012. Revisiting the relationships between text and pictures, *Children's Literature in Education*, 43(1), 4–21. https://doi.org/10.1007/s10583-011-9153-0
Skilbeck, A. 2017. Dewey on seriousness, playfulness and the role of the teacher, *Education Sciences*, 7(16), 1–14. http://dx.doi.org/10.3390/educsci7010016
Speece, M.W. and Brent, S.B. 1992. The acquisition of a mature understanding of three components of the concept of death, *Death Studies*, 16(3), 211–229. https://doi.org/10.1080/07481189208252571
Stevanović, L. 2009. *Laughing at the Funeral: Gender and Anthropology in the Greek Funerary Rites*, special editions 69, Belgrade, Serbian Academy of Sciences and Arts Institute of Ethnography.
Walter, T. 1999. *On Bereavement: The Culture of Grief*, Buckingham, Open University Press.

Part 2

Gaming Encounters in Gothic Environments

Chapter 5

Living and Dying in the City of the Damned: A Close Reading of *Mordheim's* Gothic Post-Apocalypse

Jonathan D. Stubbs

Abstract

As early as Johan Huzinga's (2016, 89–188) landmark exploration of play, *Homo Ludens* (1949), death has been recognised as integral to play. Today's digital games continue this close association. Whilst the past half-century has trended towards limiting the impact of player-death, permanent-death (*permadeath*) games provide a less-forgiving environment. As the first digital adaptation of Games Workshop's cult-classic tabletop skirmish game, *Mordheim: City of the Damned* (Rogue Factor, 2015) utilises permadeath to emphasise death's inevitability and harsh reality in the precarity of its gothic post-apocalyptic setting. Whilst the majority of apocalyptic videogames follow the comic frame, the player has no agency to overcome or change the events of Mordheim's apocalypse, setting it firmly in the gothic frame. It is substantially less about overturning disaster or saving the city, and decidedly more about looting its shattered corpse. The close reading of *Mordheim: City of the Damned*'s theme of death for this chapter identified that death and injury are simply accepted realities; ubiquitous, yet normalised. Whilst every death is significant – through permanently lost warriors – there is always another willing replacement available. Viewed alongside the warband's primary purpose – that is service to their patron – warriors' deaths not only become expected and relatively meaningless, but also financially connected. Rather than encouraging association with their warbands, players are subtly shifted to aligning with their patron, viewing the warbands and their warriors as an expendable means towards gaining digital kudos points and bragging rights amongst the other digital noble.

Keywords: Permadeath; death; Mordheim; post-Apocalypse; video games; gothic

Death, Culture & Leisure: Playing Dead, 69–82

doi:10.1108/978-1-83909-037-020201012

Introduction

Death has long been a common feature of play. Johan Huzinga, in publication *Homo Ludens* (1949), recognises the integral relationship between death and play, from mock battles seen in medieval jousting and Renaissance duels, to the penalty for failure in ancient riddle challenges (Huzinga, 2016, 89–135). Tabletop games regularly incorporate death as a means of scorekeeping, as in *Risk*; and as a key narrative feature, as in *Cluedo*. Digital games continue this close association, from the action-hero antics of first-person shooters (FPSs) such as *Call of Duty 2* (Infinity Ward, 2005), to the evocative environment of decay in *Fallout 3* (Bethesda Game Studios, 2008), and the mythical figure of Death as a non-player character (NPC) in *The Sims* (Maxis, 2000).

In the early days of coin-op arcades, death served to mark and limit progress, requiring players to spend their money to try again (Wenz, 2014, 311). As digital games moved from the arcade to the home death became increasingly less significant, with manufacturers seeking to keep players engaged for longer to justify the substantial increase in upfront, and potentially ongoing, cost for the player. Since the mid-1990s, digital games have evolved to include elaborate autosaves, checkpoints, and manual saves, providing the player 'an opportunity to retry … the effective cost' being but 'lost time, postponing the completion of the game' (Juul, 2013, 83). However, the spectre of death has returned to some games through the reintegration of permanent death mechanics (known as *permadeath*), once again providing consequence to in-game death (Frome, 2016, 160–161). Whilst this enables game designers to offer more challenging experiences, it also provides a reflection on the increased sense of precarity within the games industry, and culture at large, in the wake of the 2008 global financial crisis (Chang et al., 2017, 104–107).

Rogue Factor's *Mordheim: City of the Damned* (Rogue Factor, 2015) is one such example of the recent trend in *permadeath* games. This chapter considers death in *Mordheim's* single player Mercenary Warband campaign, drawing on the close reading strategy proposed by Jim Bizzocchi and Joshua Tanenbaum (2012), in their analysis of *Mass Effect 2*'s (BioWare, 2010) narrative. The resultant strategy employed multiple battleground missions and story-act playthroughs, extensive recording of contemporaneous notes and screenshots, and the use of supplemental *Let's Plays* to provide additional story-act data (Bizzocchi and Tanenbaum, 2012, 395–396). The close reading focussed on aspects of narration (including voice-over notifications, mission briefings, and title menu audio), game-world environment (including interactive menu screens and maps, both battleground missions and story-acts) and post-battle sequences (including post-battle reports, injury rolls and daily reports). Several days of notetaking and play, produced some 30 pages of hand-written notes and around 150 screenshots, adding to almost 650 hours of pre-existing play.

Welcome to Mordheim: City of the Damned

Produced by developer Rogue Factor; *Mordheim: City of the Damned* (subsequently *Mordheim*) was released on PC in 2015 and ported to Xbox One and

PlayStation 4 a year later. The first digital adaptation of Games Workshop's cult-classic tabletop skirmish game (Games Workshop, 1999) captures its analogue forebear's evocative feel; blending turn-based tactical gameplay, unit customisation and roleplaying game (RPG) elements. The player leads one of six warbands, customising, equipping and guiding them through Mordheim's ruined streets. Four warbands were available at release – Human Mercenaries (warriors drawn from the Imperial province of Reikland), Sisters of Sigmar (an exclusively female Cult of Sigmar sect, whose centre of worship lies within Mordheim), Skaven of Clan Eshin (humanoid rat-creatures that occupy the dark and hidden places of the Old World), and the Cult of the Possessed (degenerate fanatics and mutated horrors who worship the Chaos gods). The other two were added in separate downloadable content (DLC) packs – Witch Hunters (or Templars of Sigmar, zealous fanatics come to root out evil with sword and fire), and the Undead (zombies, ghouls, vampires, and necromancers, the unholy vassals of Count Vlad von Carstein). Consisting of a Leader, Heroes and Henchmen; the warband gains in strength and ability as it earns experience, picking up and purchasing new equipment and unlocking new warrior options as it progresses.

Single-player campaign missions can be undertaken each day of a warband's journey (the way in which the progression of time is managed throughout the game) – choosing from a random selection of battleground missions or undertaking their current, warband-specific story-act. Battleground missions are common to all factions, pitting the player's warband against a randomly generated, computer-controlled rival. It is primarily through these procedurally generated encounters that warbands earn experience, take from their fallen foes, and gather gold and other resources. Whilst victory provides greater rewards, defeat does not impede progression. Each day provides a new selection of potential battleground missions and new opportunities.

Meanwhile, the story-acts form a warband's individual eight-part narrative, divided into two chapters of their faction's campaign. Unlike the procedurally generated battleground missions the story-acts are largely pre-scripted, allowing for elaborately designed environments and engaging set opponents, whilst presenting specific challenges and orders from their patron. Each story-act features pre- and post-script narratives delivered by their faction's Dramatis Personae, the named character who accompanies the warband during story-acts and forms the primary narrative connection between warband and patron. Unlike battleground missions, each story-act must be undertaken sequentially. Success is rewarded with gold and equipment from a satisfied patron, and the promise of a new mission to come. Defeat requires the story-act be undertaken again, however, whilst this impedes a warband's story-act progression their *Mordheim* journey continues unabated.

Warbands also receive regular, time-limited, smuggling requests from their patron for wyrdstone shipments – 'strange green-black stones ... discovered' in ruined Mordheim, whose 'uncanny properties could work wonders' (Rogue Factor, 2015). Regardless of a warband's present condition, or their story-act progress, wyrdstone shipment requests persist and their deadline draws nearer. Successfully fulfilling a request grants financial reward, however a request for a larger shipment follows the brief, three-day respite. Thus, the warband journeys through these twin narratives – each serving different demands of their patron.

Mordheim employs a *permadeath* mechanic, similar to *XCOM: Enemy Unknown's* (Firaxis Games, 2012) Ironman difficulty. However, it also employs a singular scaling difficulty rather than traditional difficulty levels, balancing battleground mission opponents to the warband's current rating (a grading metric based on experience, skills, equipment, injuries, etc.). As such, *permadeath* is always in effect, rather than serving as a higher-level optional challenge for more experienced players – something that has provoked some criticism, with one reviewer noting that 'Mordheim is unapologetically hardcore' (Andrews, 2016). This is achieved through assigning each created warband a single save-file, which is continuously overwritten through a permanently active autosave feature which triggers after all actions, both in-mission and in the warband management screens. The effect precludes the possibility of reloading or *save-scumming*, the practice of saving before a prominent or challenging sequence and reloading in the hope of an improved outcome (Johnson, 2018, 207–210). When a warrior is taken out of action they 'might lose a limb ... or worse ... in Mordheim ... a dead unit is lost forever!' (Mordheim: City of the Damned, n.d.). The implementation of these mechanics produces an appropriate sense of risk and fear, for its gothic post-apocalyptic setting.

Gothic Post-apocalypse

The concept of gothic post-apocalypse draws on the idea of the gothic apocalypse outlined in Susan Tyburski (2013). Identifying recent trends in cinematic eco-apocalypses, Tyburski builds on the two apocalyptic frames identified by Stephen O'Leary – the life affirming (or comic) and the destructive (or tragic – alternatively termed horror or gothic, by Tyburski (2013, 154) and others. One principal distinction between these two frames is the question of human agency. According to O'Leary the comic apocalypse is contingent on 'human choice and activity in the world' and evil is 'something to be overcome', whereas the tragic 'views this End as predestined and catastrophic; evil ... can only be overcome by divine intervention rather than human action' (O'Leary, 1984, 85). Therefore, in the comedic apocalypse, the dread and horror of destruction is ultimately a challenge defeated with the right application of brains, brawn and skill, the hero able to rewrite their own 'destiny in a way that affirms the continuation of human life' (Tyburski, 2013, 154). By contrast, the gothic apocalypse is typically more pessimistic and cynical in tone, eschewing any easy fixes, heroic reversals or the action movie trope of success through good old American ingenuity. Instead, the hero is powerless to alter the ultimate outcome, falling short of the oft anticipated hopeful note of civilisation rebuilding from the ashes (O'Leary, 1984, 68).

As Greg Garrett notes this 'archetypal narrative ... of the apocalypse ... has been told and retold in the media of television, films, comics and graphic novels, and video games' (Garrett, 2010, 140). Rachel Wagner (2012, 187) also identifies a continuing fascination with apocalyptic imagery, suggesting that, whilst perhaps more commonly associated with (religious) prophecies of end-times destruction, 'video game worlds are surprisingly rife with apocalyptic imagery'. As with other media, apocalyptic imagery finds particularly comfortable and common expression in the genres of science fiction, fantasy and horror (including the gothic). Like the big/small screen apocalyptic fictions, videogames that draw on apocalyptic imagery typically 'share a thematic sensibility with ancient apocalyptic in their fascination with the dualistic conflict of good versus evil' (Stevenson, 2013, 99). In situating the player within this dualistic conflict, the majority of apocalyptic videogames follow the comic frame, giving the player the sense of agency within the narrative to overturn, postpone or escape the apocalypse.

In keeping with its tabletop origins, *Mordheim* remains distinctly post-apocalyptic in setting. Rather than pitting the player, and their warband, against an impending threat or doom they enter the *City of the Damned* engulfed in the aftermath of cataclysmic destruction, a shattered ruin of its former grandeur. As the opening cinematic outlines:

> The fiery comet ploughed through the city gouging a deep crater in the midst of the poor quarter. The wonderous Palace of Count Steinhardt was broken. The Great Amphitheatre shattered. The Great Library blighted by a pall of ash. Fire and smoke decimated the populace …. (Rogue Factor, 2015)

These events have already happened, in keeping with *Mordheim*'s use of *permadeath* the apocalypse is not to be avoided or fought off. Much like the injuries and deaths every warband must face, the apocalyptic events described in the opening cinematic are to be accepted as the game's context. Whilst this is not particularly uncommon, given the recent resurgence in post-apocalyptic video game narratives, these titles typically remain within the comic frame, emphasising player agency in influencing this new world after the end (Schirmer, 2011, 183–186).

However, unlike post-apocalyptic games such as *Mass Effect 3* (BioWare, 2012), which situate the player in the midst of a David vs Goliath scale apocalyptic conflict, *Mordheim* is not an under-dog rebellion game. Just as the apocalyptic events cannot be avoided, they cannot be undone or reversed (Johnson-Lewis, 2013, 118–120). Unlike titles such as *Doom* (id Software, 2016), players are not inserted into a hell-scape to liberate survivors or defeat demonic or otherworldly beasts. Unlike titles such as *Half Life 2* (Valve Corporation, 2004), the objective is not to over-throw an evil dictator/power or to prevent another disaster. In *Mordheim*, the player has no agency to overcome or change the events of the apocalypse, setting it firmly in the gothic frame. It is substantially less about overturning disaster or saving the city, and decidedly more about looting its shattered corpse. Even the pious Sisters of Sigmar are not on a mission to save ruined Mordheim, but to atone for their own perceived state of fallen, sinfulness – by hording its wealth and the wyrdstone in their own vaults for their own purposes.

Through its aesthetic, and perpetual sense of dread, *Mordheim* is also a distinctly gothic game. From the player's first steps into the game-world, they encounter the gothic. Sequential menu screens draw the player down the desolated twin-moonlit street, lined with dead trees swaying in a soundless breeze, through the gates and into the city. Once distantly illuminated against the night's sky by a faint ghostly green glow, the player is drawn into the city and the courtyard that holds the warband's camp – decay evident in the dimly lit tattered banners and crumbling stonework. The gothic sense of the uncanny can be seen in the mutated horrors and warped environment, set alongside the abandoned homes and signs of the familiar domestic made unhomely and the macabre displays of corpses staged as a mockery of their former life (Partridge, 2016, 132–133). A sense of horror and disgust is evoked by the thematically bubbling blood geysers springing from the cobbled streets, prominent blood stains and the noxious emissions of the corpulent mutation growths clinging to buildings, whose eyes follow the warriors around the battlefield. The supernatural is regularly drawn on in the procedurally generated maps used for battleground missions, with impassable walls of spectral, green apparitions circumscribing the arena of play where natural barriers are absent. Meanwhile, dismembered body parts, decorated skulls, and displayed corpses dot every interactive menu screen and map, with flaming corpses serving as occasional thematic light sources.

Therefore, *Mordheim* is both distinctly gothic and distinctly post-apocalyptic. Thus, this can be tentatively viewed as an expansion on Tyburski's earlier concept of the gothic apocalypse, effectively a sub-genre, the gothic post-apocalypse. Whilst more extensive study would be required to define such an actual sub-genre, it aids in framing this project's exploration of the theme of (perma)death in *Mordheim: City of the Damned.*

Life and (Perma)death in *Mordheim*

Death looms large in *Mordheim*'s gothic post-apocalyptic world, from narration and macabre corpse presentation to ever-present, yet normalised threat. Having established *Mordheim's* gothic post-apocalyptic framing, the remainder of this chapter turns to, (1) consider the thematically harsh realities of life within ruined Mordheim and the impact of death's ubiquity; (2) explore the death mechanic as a bleak reflection of *Mordheim*'s gothic post-apocalyptic setting; and (3) analyse the implementation of *permadeath* and *Mordheim's* subsequent engagement with the concept of precarity.

Life

In beginning, a new Mercenary warband for this project the title menu's opening voiceover – 'Will this day be your last?' (Rogue Factor, 2015) – heralded the grim prospect of those braving Mordheim's ruins. As already noted, the battleground mission maps feature gothic aesthetics commonly associated with death – from the city's burnt-out, decaying ruins to the mutated growths and decoratively displayed corpses. The abandoned remains of Mordheim's former residents and unfortunate treasure hunters often serve as valuable, if morbid, loot sites. Elsewhere, corpses are formed into grotesque displays, for example a pair of jousting knights, hung from balconies, crucified on a building's damaged beams, strapped to abandoned carts and adorning ruined buildings. Meanwhile, the pseudo-Medieval architecture often prominently displays skeletal designs, skulls with rictus grins and effigies of the robed figure of Death or the winged angel of death.

Death is not only prominent in *Mordheim*'s aesthetic, it is also unavoidable in its gameplay experience – which Matthew Spokes terms implicit and explicit encounters with death (Spokes, 2018, 137–139). For a warband to develop, earn experience, pick up treasures and gather the necessary wyrdstone they must battle Mordheim's other treasure hunters and assorted horrors. Whilst differing mission types exist for the battleground missions, each providing its own secondary objective (Marked for Death – eliminate specific opposing warband members; Wyrdstone Hunt – gather a certain percentage of available wyrdstone; and Crush Their Will – capture the opposing warband's banner), the primary objective remains the same – 'take the enemy out of action'. Regardless of a battleground mission's secondary objective, victory and defeat are determined by which warband routs first – a condition contingent on the elimination of warband members. Victory lies in forcing the enemy to rout, whilst defeat results from being routed.

Win or lose, out of action warriors face the uncertainty of the post-battle injury roll which determines the extent of their injuries; the unavoidable consequences of *Mordheim*'s *permadeath* mechanic. From the inexperienced, novice henchmen to the trained and seasoned hero, no warrior is spared this threat, not even the warband's leader (the nominal player-character). Warriors enduring an injury roll may survive unscathed, but risk suffering disability (e.g. severed leg/arm and loss of an eye), mutilation (e.g. scarring and burst eardrum), mental scarring (e.g. stupidity and megalomania), and even death. Whereas, in other games, players may restart or quit a faltering mission, abandoning the battleground is rarely advisable since retreating effectively puts all the player's warriors out of action – subjecting them to the resultant injury roll.

Therefore, death (and crippling injury) is not only an accepted reality within the world of *Mordheim*, it is a ubiquitous, yet normalised, threat. Whether through poor decisions, bad luck or the vagaries of the Random Number Generator (RNG – a game mechanic designed to simulate random chance) every warband will face injury and death. Even the Undead warband is not immune to this threat. In *Mordheim*, zombies, ghouls and vampires are just as susceptible to losing a limb, suffering mental scaring or dying as a result of an injury roll as any human or Skaven warrior.

Death is not only ubiquitous across the *Mordheim* aesthetic and experience, it is also varied. Whilst individual warriors are subject to risk of death through their post-battle injury roll, they can also suffer death of utility, and death through neglect. Death of utility comes through injury and disability. Each injury received applies potential penalties, ranging from reduced speed and agility from a severed leg to diminished accuracy caused by the loss of an eye, and occasional paralysis due to the mental condition *stupidity*. In addition to the impact of a permanent injury, their healing costs in both finances and time – temporarily rendering the warrior inaccessible. Therefore, the player must choose between healing and maintaining injured warriors with diminished combat and resource gathering potential, or firing them in favour of new, healthy hired swords[1] – an act rewarded with an achievement when played through Steam; Cold-Hearted. Death through neglect describes the process of allowing injured warriors to succumb to their injuries, ignoring the warnings and notifications received in the daily report, and refusing to pay for their treatment – a practice rewarded with an achievement when played on Steam; Heartless (TrueSteamAchievements, n.d.).

The *Dramatis Persona* alone is sparred the ignominy of such deaths, or the risk of disabling injury. Accompanying warbands only during their story-acts, *Dramatis Persona* do not form part of the warband's regular roster and therefore are neither subject to post-battle injury rolls, nor may they be dismissed. *Dramatis Persona* are further distinguished from a warband's regular warriors, as their battlefield survival is uniquely linked to the victory conditions (barring a handful of mission-objective characters in later story-acts). Whilst the *Dramatis Persona's* elimination results in automatic defeat, they return unscathed for the warband's next attempt, however, any warriors taken out of action still face the post-battle injury roll.

Death

The threat of death is not unique to *Mordheim*, but, as John W. Borchert notes, death 'has been an imminent threat ... since the earliest cabinet arcade games' (Borchert, 2018, 1). This disruption in the player experience, the marking of failure, poor luck, or bad timing is presented in myriad ways. For example, the dramatic out-of-body kill-cam sequences depicting 'the player's death from the killer's perspective' (Adams, 2014) in *Sniper Elite: V2* (Rebellion Developments, 2012). Meanwhile, *The Elder Scrolls V: Skyrim* (Bethesda Game Studios, 2011) utilises the spectacle of un-skippable, forced-third-person visuals, sensationalising player-death with 'moribund digital imagery' (Barnes, 2018, 112). Similarly, *Max Payne* (Remedy Entertainment, 2001) emphasised the event of player-death through third-person, slow-motion sequences, centring around the dying protagonist, which *Outlast* (Red Barrels, 2013) further enhanced through fading to stark monochrome, as the player-character's life ebbs out. Whereas, survival-horror games, known for their vivid depictions of blood, gore and dismemberment, such as *The Evil Within* (Tango Gameworks, 2015), revel in ever more gratuitous cinematic demises. However, in keeping with the bleak and unforgiving gothic post-apocalyptic environment of *Mordheim*, death lacks any great sense of ceremony, regardless of the ultimate manner – result of an injury roll, death of utility, or death through neglect.

---

[1]'Hired swords' refers to a varied and changing collection of low-mid level warriors available to hire, complete with their appropriate attribute and skill points, though at a substantially increased price compared to untrained recruits.

Death through neglect only warrants a brief note in the warband's daily report, informing the player that, having ignored the numerous warnings, their warrior has died. The short, two-line message is easily missed or ignored, sitting alongside various other updates on the health and training of warband members. There is nothing distinct to mark out the occurrence as bearing any special significance, no voiceover or auditory cues announce their passing. Henchman and hero alike, every death through neglect is essentially unremarkable, noting only that '[their] injuries were left untreated for too long' (Rogue Factor, 2015).

Death as the result of an injury roll receives the greatest attention; however, even this is considerably subdued when compared with many contemporary titles. Warriors suffering this fate merely fall to the ground with an appropriate final vocalisation signalling their demise. A brief note displaying the fact of their death and the other warband member's brief recognition of their passing (where other injury's description and impact would be listed) is accompanied by an appropriately underwhelming voiceover – for example 'Another one bites the dust' (Rogue Factor, 2015). Death of utility, on the other hand, is essentially a non-event. Much like the sale of excess or useless equipment, the crippled warrior is dispatched with a simple command, discharging them from the warband and leaving them to their own fate.

The unceremonious manner in which the irreversible death of even valued and skilled warriors is treated has the effect of diminishing its importance, a prospect that even the warband leader (the nominal player-character) is subject to. The general sense of disregard shown towards a warrior's death echoes the very matter-of-fact style of many roguelikes. Eschewing the spectacular visuals and visceral cinematics players are commonly treated to, roguelikes are known for their simple memorialisation of death, typically noting its occurrence and providing a brief record of achievement (Parker, 2017, 132–136). Whilst death in roguelikes is usually synonymous with the end of a player-character's run, requiring some degree of restart as in *Enter the Gungeon* (Dodge Roll, 2016), *Mordheim*'s use of warbands limits its impact to the particular warrior. However, where comparable warband/squad-based *permadeath*-games might include a lasting memorial to fallen warriors, as with *Darkest Dungeon*'s (Red Hook Studios, 2016) 'Graveyard' or *XCOM: Enemy Unknown*'s (Firaxis Games, 2012) 'Memorial Wall', those who perish in *Mordheim* are denied even this basic recognition. Their death is unremarkable, their utility has been spent, eager replacements await and the warband's journey continues.

Permadeath and Precarity

As a *permadeath* game, *Mordheim* is noted for having an arguably steep and unforgiving learning curve. This is not uncommon with *permadeath* games, which typically seek to connect serious consequences to in-game death, reversing the trend of viewing death as 'temporary setbacks to in-game progress' (Chang et al., 2017, 104). Due to its permanently active autosave, every death and crippling injury sustained is irreversible. The consequences of poor decisions, unfavourable RNG results, or overwhelming opponents cannot be avoided, nor does an earlier save-point exist to enable the player to try again. Therefore, every warrior's death bears a certain significance. They cannot be respawned or reloaded. They are simply removed from the warband's roster. Whilst their equipment returns to the warband's supplies, any gold and time invested in their training is obviously lost.

However, as with many modern roguelikes and other games utilising *permadeath* mechanics, *Mordheim* employs a variety of features aimed at 'mitigating the punitive aspects of *permadeath* ... [and] making the complexity ... a little more user friendly' (Parker, 2017, 136). Many of the features used would be familiar to players of FPSs and RPGs. For example, tutorials, commonly incorporated into a prologue or initial training level in FPSs, may be undertaken on *XCOM: Enemy Unknown*'s (Firaxis Games, 2012) Ironman mode. Meanwhile, the use of player-linked rather than run-through linked unlockable perks – characters, equipment, secret rooms – provide a sense of achievement and progress as with *Enter the Gungeon*'s (Dodge Roll, 2016) rescuable NPCs; providing the player with new equipment possibilities. Alternatively, despite incorporating a *permadeath* mechanic, the use of different difficulty settings can enable the player to influence the level of challenge, as in *Darkest Dungeon* (Red Hook Studios, 2016). Additionally, warband/squad-based *permadeath*-games often allow for the purchase of replacement warriors and expanding rosters, as in *Mordheim*. Therefore, whilst every death bears significance in lost resources, there is always another warm body available to take their place.

During the close reading for this chapter, a half-dozen warriors were unceremoniously replaced by the tenth mission, through injury and death, including the warband's leader. When pertaining to lower level warriors, replacement rather than maintenance is often the more advantageous choice and is easily justifiable – regardless of whether they're an inexpensive henchman or a more valuable hero. Such warriors have typically received no costly training, the expenditure on them has been minimal and a comparable replacement is relatively inexpensive. It may even be advantageous to replace mid-level warriors, in whom time and resources have been invested, since new hired swords may apply all available skill points and perks, without incurring the normal costs and unavailability whilst training. A prospect made more tantalising by the provision of free-to-hire, mid-high-level hired swords as higher tier rewards for wyrdstone smuggling – both from the warband's patron and two other optional recipients of excess wyrdstone.

Meanwhile, the Veteran Skills system unlocks player-specific rewards, which can be applied to later warbands – including additional starting gold and various perks providing discounts on warband purchases. Whilst such features are not uncommon in modern *permadeath* games, designed to mitigate the impact of a substantial defeat and the need to start over, its execution in *Mordheim* potentially encourages the wholesale sacrifice of earlier warbands for the benefit of later ones. Additionally, once the player reaches Veteran Skills rank five all new warbands may begin at rank five with all the relevant benefits, effectively mitigating the loss of entire warbands. The combination of player-specific bonuses and regular availability of hired swords has the effect of limiting the impact, particularly to long-standing players, of what would otherwise be a harsh *permadeath* mechanic. When combined with the lack of significance attached to a warrior's demise, death can become effectively meaningless – as even the warband leader is relatively easily replaced.

When considered in light of the warband's primary purpose in ruined Mordheim – outlined to each warband as service to their patron – warriors' deaths not only become expected and relatively meaningless, but also financially connected. This is particularly evident in the patron's regular demands for ever-increasing wyrdstone shipments, a narrative that undercuts everything as wyrdstone collection becomes an un-written primary goal extending beyond the conclusion of the story-acts. The Mercenary campaign emphasises its centrality, stating that '[sending] wyrdstone to Baron von Leitdorfer will give their patron an advantage over the other nobles' (Rogue Factor, 2015). With the deadline prominently ticking down throughout the war-camp menus, the player is pressed on to hunt this precious mineral, potentially sacrificing warriors and firing injured ones rather than risk inactivity whilst they heal.

Wyrdstone collection's centrality is unavoidable. Regardless of the warband's condition, shipment request deadlines continue to tick-down, whilst their successful fulfilment grants only a short reprieve before a new, larger shipment request arrives. Failure to fulfil a shipment request results in a demand for a greater quantity in a shorter time, with the warning that if 'you are incapable of the job, there are plenty of warriors to take your place' (Rogue Factor, 2015). Subsequent failures are not treated so lightly by the Baron, escalating to inflicting disabling injuries before mysteriously disappearing a random warrior; 'Failure has consequences' (Rogue Factor, 2015). In their patron's continued pursuit of wyrdstone, every warrior is expendable, including the leader.

Further, aside from a warband becoming essentially unusable through excessive injury or lack of resources, failure to fulfil shipments is the only game over condition. Even successful completion of the warband's eight-part story-act campaign does not end their journey, despite being informed their recent endeavours 'should settle … [the Baron's] rivalry with the Duke once and for all' (Rogue Factor, 2015). Instead of a grand cinematic experience or neat narrative conclusion the player receives the customary 'Mission Reward' notification two days later, stating the Baron 'has granted us a reward to ensure our continued loyalty' (Rogue Factor, 2015). Rather than a well-deserved rest, the warband receives some gold and a handful of trinkets before being encouraged to continue gathering wyrdstone for Baron von Leitdorfer – shipment requests and all. Once again, the player is reminded that the warband's primary duty is not saving the city, preventing a second disaster or vanquishing the hidden evil behind the comet-strike. Instead, they serve at the behest of a distant master, buying their loyalty with tokens, punishing failure and judging them as ultimately expendable in the pursuit of increasing their patron's wealth, renown, and status.

In this way, *Mordheim* connects *permadeath* with the concept of precarity. Each warriors' value lies in their capacity to serve their patron, failure's consequences are harsh and unavoidable, and willing replacements abound. Whilst such conditions might feasibly encourage a sense of comradeship pitting player and warband against the demanding, uncaring patron, the game's implementation of *permadeath* serves to distance the player. Despite each warband's opening narrative identifying the player with the warband leader, their expendability disrupts this illusion. Meanwhile, the threat of permanent disabling injury and the unceasing need to acquire wyrdstone encourage an attitude as unforgiving as the gothic post-apocalyptic world the warband occupies, encouraging the player to treat even the leader (their nominal avatar) with the same financially motivated dispassion as the Baron – distant, detached, and disembodied. This is further engendered through the awarding of achievements for meeting in-game criteria at the expense of the warband, including 'Triage' ('Have 6 injured warriors at once') and 'Bruised and Beaten' ('Suffer 500 permanent injuries'), or, for sacrificing an entire warband, 'Angry Boss' ('Game Over after 4 failed delivery') (TrueSteamAchievements, n.d.). Rather than encouraging association with their warbands, players are subtly shifted to aligning with their patron, viewing the warbands and their warriors as an expendable means towards gaining digital kudos points and bragging rights amongst the other digital nobles.

Conclusion

As a gothic post-apocalyptic game, death features heavily throughout *Mordheim*. It provides a rich background and evocatively gruesome world for the player to explore, from its interactive menus and narration, to the elaborate maps and detailed warband design. Featuring a fairly unforgiving *permadeath* mechanic, the player is invited to guide their warband through the ruined city, seeking fortune and glory in service to their demanding patron. As part of a growing trend of *permadeath* games, *Mordheim*'s motif of death emphasises its inevitability and harsh reality in the precarity of this gothic post-apocalypse. Whilst each death is irreversible and arguably more significant, the institution of various mitigation mechanics drastically reduces death's impact. Whereas other titles might valorise the fallen, *Mordheim* delivers an unceremonious treatment of death, even rewarding a somewhat callous disregard for warband members with the provision of player-based achievements. Ultimately, the message is one of meaninglessness and the expendability of life in the pursuit of wealth for the warband's distant patron, and success for the player. In *Mordheim,* they are all just part of the 'endless pilgrimage of the greedy and ambitious, fighting and dying in a place now called: The City of the Damned' (Rouge Factor, 2015).

Interactions and Sensoria

Dodge, R. 2016. *Enter the Gungeon*, Devolver Digital.
Firaxis Games. 2012. *XCOM: Enemy Unknown*, 2K Games and Feral Interactive.
Nicalis. 2014. *The Binding of Isaac: Rebirth*, Nicalis.
Red Hook Studios. 2016. *Darkest Dungeon*, Merge Games.
Rodeo Games and Twist Play. 2015. *Warhammer Quest*, Chilled Mouse.

References

Adams, M.B. 2014. Spectacular mortality: intersections of punitive and educational player-death, *First Person Scholar*, 7 May. Available at: http://bit.ly/2WoBOdp

Andrews, S. 2016. Mordheim: city of the damned (console version) review. *Trusted Reviews*, 7 May. Available at: http://bit.ly/31THddN

Barnes, R. 2018. Perilous and peril-less gaming: representations of death in Nintendo's wolf link Amiibo. *Gamevironments*, 9, 107–134.

Bethesda Game Studios. 2008. *Fallout 3*, Bethesda Softworks.

Bethesda Game Studios. 2011. *The Elder Scrolls V: Skyrim*, Bethesda Softworks.

BioWare. 2010. *Mass Effect 2*. Electronic Arts.

BioWare. 2012. *Mass Effect 3*. Electronic Arts.

Bizzocchi, J. and Tanenbaum, J. 2012. Mass effect 2: a case study in the design of game narrative, *Bulletin of Science, Technology and Society*, 32(5), 393–404. https://doi.org/10.1177%2F0270467612463796

Borchert, J. W. (2018). Introduction to a special issue on video gaming and death, *Gamevironments, 9*, 1-11. Retrieved from http://nbn-resolving.de/urn:nbn:de:gbv:46-00106987-14

Chang, A., Constantino, J. and Soderman, B. 2017. The multiple lives of permadeath: an introduction, *Journal of Gaming and Virtual Worlds*, 9(20), 103–121. https://doi.org/10.1386/jgvw.9.2.103_1

Dodge Roll. 2016. *Enter the Gungeon*, Devolver Digital.

Firaxis Games. 2012. *XCOM: Enemy Unknown*, 2K Games and Feral Interactive.

Frome, J. 2016. Video game sadness from planetfall to passage. In *Video Games and the Mind: Essays on Cognition, Affect and Emotion*, Eds B. Perron and F. Schröter, pp. 158–173, Jefferson, MacFarland & Company Inc. Publishers.

Games Workshop. 1999. *Mordheim*, Games Workshop.

Garrett, G. 2010. 'I saw one like a son of man': the eschatological saviour in contemporary film. In *Reel Revelations: Apocalypse and Film*, Eds J. Walliss and L. Quinby, pp. 140–154, Sheffield, Sheffield Phoenix Press.

Huzinga, J. 2016. *Homo Ludens: A Study of the Play-Element in Culture*, Kettering, Angelico Press.

id Software. 2016. *Doom*, Bethesda Softworks.

Infinity Ward. 2005. *Call of Duty 2*, Activision.

Johnson, M.R. 2018. *The Unpredictability of Gameplay*, New York, NY, Bloomsbury Publishing.

Johnson-Lewis, E. 2013. After the end: post-apocalyptic narratives in Lost, Jericho, and Battlestar Galactica. In *Small Screen Revelations: Apocalypse in Contemporary Television*, Eds J. Aston and J. Walliss, pp. 118–132, Sheffield, Sheffield Phoenix Press.

Juul, J. 2013. *The Art of Failure: An Essay on the Pain of Playing Video Games*, Cambridge, MA, The MIT Press.

Maxis. 2000. *The Sims*, Electronic Arts.

Mordheim: City of the Damned. n.d. Presentation. *Mordheim: City of the Damned*, 11 April. Available at: http://bit.ly/31VxSSC

O'Leary, S.D. 1984. *Arguing the Apocalypse: A Theory of Millennial Rhetoric*, Oxford, Oxford University Press.

Parker, R. 2017. The culture of permadeath: Roguelikes and terror management theory, *Journal of Gaming and Virtual Worlds*, 9(2), 123–141. https://doi.org/10.1386/jgvw.9.2.123_1

Partridge, C. 2016. Temenos lecture 2015: death, the gothic, and popular music: some reflections on why popular music matters, *Temenos – Nordic Journal of Comparative Religion*, 52(1), 127–150.

Rebellion Developments. 2012. *Sniper Elite: V2*, Rebellion Developments.

Red Barrels. 2013. *Outlast*, Red Barrels.

Red Hook Studios. 2016. *Darkest Dungeon*, Merge Games.

Remedy Entertainment. 2001. *Max Payne*, Gathering of Developers.

Rogue Factor. 2015. *Mordheim: City of the Damned*, Focus Home Interactive.

Schirmer, J. 2011. 'We all stray from our paths sometimes': morality and survival in fallout 3. In *Network Apocalypse: Visions of the End in an Age of Internet Media*, Ed. R.G. Howard, pp. 183–199, Sheffield, Sheffield Phoenix Press.

Spokes, M. 2018. 'War ... war never changes': exploring explicit and implicit encounters with death in a post-apocalyptic gameworld, *Mortality*, 23(2), 135–150. https://doi.org/10.1080/13576275.2017.1319348

Stevenson, G. 2013. Apocalyptic war in buffy the vampire slayer, angel, supernatural and battlestar galactica. In *Small Screen Revelations: Apocalypse in Contemporary Television*, Eds J. Aston and J. Walliss, pp. 96–117, Sheffield, Sheffield Phoenix Press.

Tango Gameworks. 2015. *The Evil Within*, Bethesda Softworks.

TrueSteamAchievements. n.d. Mordheim: city of the damned. *True Steam Achievements*, 11 April. Available at: http://bit.ly/32RQ8gV

Tyburski, S.J. 2013. A gothic apocalypse: encountering the monstrous in American cinema. In *EcoGothic*, Eds A. Smith and W. Hughes, pp. 147–159, Manchester, Manchester University Press.

Valve Corporation. 2004. *Half-Life 2*, Valve Corporation.

Wagner, R. 2012. *Godwired: Religion, Ritual and Virtual Reality*, Oxon, Routledge.

Wenz, K. 2014. Death. In *The Routledge Companion to Video Game Studies*, Eds M.J.P. Wolf and B. Perron, pp. 310–316, Abingdon, Routledge.

Chapter 6

Prepare to Die: Reconceptualising Death, and the Role of Narrative Engagement in the *Dark Souls Series* (2011–2018)

Andreas Theodorou

Abstract

Dark Souls heralded a shift from the dichotomy of survival horror, and instead, thrust the player into a world where narrative was everywhere (if only you dared to look). This chapter explores the reimagination of Gothic narrative and narrative engagement in the cryptic and fragmented nested narratives of the iconic FromSoftware, Inc. series. In doing so, this chapter highlights the emergence of a hybrid ludo-narrative form within the Gothic genre, and examines the ways in which the series presents said narratives to the player as it shifts the onus of narrative engagement from the storyteller to the one now living the experience. The chapter explores video-ludic interpretations of death, play, and experientiality through the lens of video game studies, and posits the value of the series as a defining moment in the Japanese action role-playing game genre.

Keywords: Narrative; gothic; dark souls; video-ludic; death; play

Introduction

On 22 September 2011, FromSoftware released *Dark Souls*, a game which drew heavily from its spiritual predecessor, *Demon's Souls* (FromSoftware, 2009). *Dark Souls* quickly caught the attention of gamers, which lead to numerous critical recognitions.[1] *Dark Souls*' most notable feature was the sheer lack of difficulty

[1]Including: Electronic Gaming Monthly, Game Revolutions, IncGamers, and Q-Games 'game of the year'; GameTrailers' 'best role-playing game'; Edge's 'best game of the generation'; USgamer's 'best game since 2000; and two nominations for BAFTA games award for 'best strategy game'.

Death, Culture & Leisure: Playing Dead, 83–94

doi:10.1108/978-1-83909-037-020201013

settings – a novel and welcome decision that deviated from the near-universal norm of in-game tutorials which slowly introduced players to the process of gameplay as well as allowing players the autonomy to decide on the level of challenge they wished to encounter (Coward, 2018). Instead, *Dark Souls* thrust players headfirst into a world in which death was not only constant, but reimagined. Death, in this instance, is coupled with beautiful cinematics, as well as cryptic and fragmented nested narratives (Elsaesser, 2009; Kroener, 2014). This chapter examines the reconceptualisation of death and the onus placed upon players engaging with the narrative of the *Dark Souls* trilogy (FromSoftware, 2011, 2014, 2016, 2018). Through a content analysis of the franchise, I contend that, through its unforgiving and grotesque world, the series forces the player to engage with, and reframe their approach to death via a ludo-Gothic narrative experience (challenging the gaming hegemony that to die was to fail). This chapter explores the emergence of a hybrid ludo-Gothic genre defined by the iconic narratives of the *Dark Souls* trilogy. It traverses the dissemination of fragmented and cryptic nested narratives, and addresses the shift in the narrative form of Gothic video games. In closely analysing the modes in which narrative is conveyed in *Dark Souls* (both written and sensorial), and the way in which the player is encouraged to seek out narrative understanding, this chapter examines how global Gothic tropes are presented in the written aspects of the game, the visual aesthetics, sounds, motions, and design (both character and landscape). This chapter identifies a narrative of death that posits a duality of being, rather than the dichotomy previously portrayed – emphasising that the player character is neither alive nor dead, and death is not an end, but rather, a beginning (Penfold-Mounce, 2018). This chapter scrutinises the explicit discourse of death within the decaying and imploding landscapes, the *Darksign*, the painted worlds, and the cycle of linking the first flame. In order to fully address this changing narrative form, and the onus it places upon the player, I focus my analysis through the triangulation of both ludological and narratological discourses. In doing so, I build upon the works of Krzywinska and King (2006), Picard (2009), Kirkland (2009), and Perron and Wolf (2003) to highlight the shift in narrative form, and the emergence of a new, hybrid, ludo-Gothic genre.

The opening cutscene of *Dark Souls* (FromSoftware, 2011) speaks of a bloody conflict and the birth of fire. It presents a narrative of an unending struggle between life and death, wherein humanity overthrows the immortal dragon regime with the help of their 'othered' mortal abomination – reimagining mortality so as to usher in an age of (perceived) prosperity. This age, as with so many aspects of this world, is merely an illusion. For example, 'the Furtive Pygmy, so easily forgotten' (a nameless and enigmatic Lord of whom there is scant explanation), is not so much forgotten, inasmuch as they are actively obliterated from memory by the 'ruling classes'. This manipulation of history, and of the narrative we receive throughout the entirety of the game, is intentional. As Kroener (2014, 80) notes, 'we have to distinguish between mind-game film plot mechanics [...], and narrative devices that directly affect the arrangement of information within the story'. In doing so, we realise that this narrator is not reliable. While it is true that the Pygmy is so easily forgotten in this realm, the truth is that this is the result of potentially millennia of manipulation by the other Lords of Fire. This is due to the fact that, time in *Dark Souls* is not only non-linear, but also blurred and cyclical. While actions may be conducted consecutively by the player, they may, in fact, be centuries apart. Such is the flow of time in an immortal world. The cryptic and fragmented nested narratives of *Dark Souls* shroud themselves behind a non-linear timeline, and an array of unreliable information. This manipulation of information within the game, coupled with its unforgiving nature and heavily-stylised environments and soundtracks, all serve to position *Dark Souls* as a contemporary Gothic masterpiece, and thus generate 'a stronger sense of the location of gameplay within a particular, realized milieu. [Increasing] the pleasures of exploration or the extent to which a sense of presence is created' (Krzywinska and King, 2006, 124).

To further emphasise the complexity and enigmatic nature of the trilogy's narrative, we can look no further than the legacy of the Pendant. The Pendant was an unequippable item that could be selected as a gift when starting the game or found later during play. It is described in-game as: 'a simple pendant with no effect. Even so, pleasant memories are crucial to survival on arduous journey's' (*Dark Souls*). Upon the release of *Dark Souls* in 2011, lead director Hidetaka Miyazaki stated in an interview with Japan's *Famitsu* magazine that, from a roleplaying perspective, he would pick either the pendant, or nothing (McWhertor, 2012). This statement drove thousands of players to choose the pendant, in the hope of solving its mystery. However, a year later in a subsequent interview, Miyazaki noted that this item had been placed in the game with the intentions of playing a prank (McWhertor, 2012). Despite this, many players continued to dedicate time to solving the mystery of the pendant, despite the knowledge that they had been manipulated by the game designers. FromSoftware continually encouraged and goaded players into engaging with cryptic and fragmented nested narratives (Taylor, 2009). In doing so, players were forced to uncover the narrative of *Dark Souls*, rather than it being simply handed to them (as seen with the majority of AAA games). In creating such intricate narratives, FromSoftware enabled players to form unique narratological experiences that were relatable, comparable, and performable with other players (outside of the boundaries of whether or not a player has access to downloadable content or even the game itself).

These games heralded a shift from traditional horror video-games. Failing to conform to the conventional dichotomy of 'survival horror' or 'surviving horror'; since that which is dead in the environment of *Souls*, cannot die, players are unable to escape from reality itself. Instead, *Dark Souls* was the harbinger for the birth of a more broadly Gothic video-game genre, and became a gathering point for both scholars and players to engage in both narratological and ludological research and discussion – a discussion which, until recently, had been a false dichotomy of 'this over that', and one which is far too broad for the scope of this chapter (see Frasca, 2003; Murray, 2005; McManus and Feinstein, 2006; Kokonis, 2014). Kaytoue et al. (2012) first examined the growing popularity of video-game content creation and postulated that a new web community was emerging. Seven years on, Johnson and Woodcock (2017, 22–23) concluded that the rise in accessible communication media has enabled a number of people to obtain a global celebrity status – allowing us to "examine the distinctive features of digital 'cultural work'". Streamers and content creators such as VaatiVidya (Michael Samuels) and EpicNameBro (Marcus Sanders) focus almost entirely on the works of FromSoftware and the *Souls* series, a term often used in reference to FromSoftware's action roleplaying games published since *Demon's Soul* (2009). It is of little surprise that streamers concentrate on the work of FromSoftware given that Kaytoue et al. (2012) noted that *Dark Souls* was one of the twenty most popular games to be watched via livestream. VaatiVidya and EpicNameBro's digital cultural work has contributed to an ever-present, ever-growing Venn diagram of scholarly and lay analysis of video-game narrative. Furthermore, the divide between the two is bridged by publications such as *Beyond the Grave* (Mecheri and Romieu, 2017, 2018), a semi-analytical pair of books which explore the creative process behind the *Souls* series, the narratives present in the games, as well as some explicit analysis of said narratives. Such literature acts as a platform for players to better understand, and therefore appreciate, the efforts of the creators, as well as the end-product itself. In creating such platforms, developers reinforce a transnational culture that creates unique, yet shared, experiences. These narratives become a "'Trojan horse' that smuggle[s] convergence culture right into people's living rooms" (Jenkins, 2006). In making players work for their narratives, FromSoftware thus repositioned narrative as a reward, rather than a foundation. Likewise, in altering the role of death within said narratives, players were forced to revise their understanding of, and approach to, the ways in which they immersed themselves within the video-ludic Gothic – convergence culture became the prize, rather than the by-product.

The *Dark Souls* series also offers hybrid Gothic ludo-narratives through the utilisation of global Gothic tropes. Picard (2009, 98–100) observes the 'transnationality' of games, explaining that 'Japanese games are increasingly marketed for Western audiences'. As such, we can explore the deep, global intermediality of *Dark Souls* and its characters. Non-playable characters (NPCs) such as Artorias of the Abyss, and Filianore, are heavily inspired by Gothic Japanese media and literature. Artorias is shaped by the character Guts from the manga and anime *Berserk*, and Filianore is, albeit more metaphorically, drawn from the dystopian animated film 'Angel's Egg' (VaatiVidya, 2017a). Likewise, other NPCs and enemies appear to be drawn from western Gothic. For example, the character Marvellous Chester, who explains that he is a 'stranger' from another land, reflects a neo-Victorian aesthetic, that juxtaposes his more archaic and grandiose surroundings of the Royal Woods, a traversable landscape, of *Dark Souls* (2011). Interestingly in this same area, one may find Artorias, who lingers in a coliseum-like structure not far away. As if highlighting the combination of Gothic influences, here we find two imposing figures, symbols of the transnationality of contemporary Gothic ludo-narrative. We can further see this hybridity in examining characters such as the Chaos Witch Quelaag. As a 'daughter of chaos', Quelaag represents the combination of Japanese supernatural folklore and western horror. Quelaag appears to be a combination of *Yūrei* (幽霊) and *Yōkai* (妖怪), with her pale-white skin, long black hair, and demonic spider-like lower-body. The colour white and its symbolism are commonplace in the Gothic genre, and *Dark Souls* is no different. In Shinto, the colour white is a symbol of purity, thus, the contrast of her pale white skin and jet-black hair, coupled with her lower-spider-half, render her a grotesque amalgamation of Japanese Gothic folklore. Alongside this, we can further associate the colour white with Japanese funeral rites, and the white cloths which decorate funeral altars (Shinya, 2004). Yet, while her appearance is heavily rooted in Japanese Gothic, her 'birth' appears to be inherently western. Quelaag's present form is the result of a failed attempt to replicate the first flame, spewing forth demonic creatures and distorting the beings around, what would come to be known as, the Bed of Chaos. Characters such as Quelaag further the fetishisation of Japanese cultures as a marketable, consumable subject. In combining Japanese folklore with western narrative and fetishism, *Dark Souls* offers global hybrid Gothic ludo-narratives.

The recurring theme of death in the *Souls* series is further projected in the decaying landscapes it presents. Throughout the series, players are continually exposed to decrepit and toxic environments that serve to desensitise them; gradually making death a more explicit and constant presence in the ludo-Gothic experientiality of the series. This progresses throughout the series until the player finds themselves at the Kiln of the First Flame in *Dark Souls III*. At this point in the series, the cycle of death has, presumably, continued for millennia; constantly distorting the world to such an extent that it is now collapsing in on itself. This narrative of decay continues until little but dust and ash remain at the end of the Ringed City DLC, and the end of time in the realm of Lordran (*Dark Souls III*). With this in mind, there is far more for the player to experience narratologically than just the item descriptions and NPCs; the created world of *Dark Souls* demands the player's attention as well. Interaction with the landscape is not only encouraged by the game, but at points, is essential for a player's progression. Bonfires, treasures, and locations are casually hidden behind illusory walls with the encouragement for players to support other players. 'Hit here' text-boxes (deposited by players with in game methods of communications) allowed for the uninitiated to benefit from the experiences of others, and 'try jumping' signs, more often than not, led to suspicion, and later derision from players, as they would plummet to their deaths; having been deceived by their more sadistic peers, sometimes referred to as the 'baptism of fire' by 'griefing' players (Hunter et al., 2015). From a ludological perspective, the environment and landscapes of *Dark Souls* try to continually deprecate the player. Black (2016) examines the role of verticality and reward in *Dark Souls* in his discussion of the importance of Blighttown, a dark, chaotic, and brutal environment that serves as a definitive example of the game's approach to effort and reward. It is also considered as one of the most unforgiving spaces in the world of *Dark Souls*. Black (2016) argues that, in *Dark Souls*, verticality and reward are used as 'a means to make the player feel powerful, even for a brief time', and as such, the descent into Blighttown and The Depths represent a hyperrealist approach to reward and effort, wherein, as with real life, you can put in a great deal of effort with minimal to no reward. He notes that in Blighttown, the game 'does not elevate you to a position of relative importance', but seeks to reiterate your insignificance, and 'teach you to appreciate the little victories' (Black, 2016). This constant, albeit brutal, interaction with the world around them forces player engagement with non-explicit narratives, allowing them to form their own, unique interpretations.

This constant needs to interact and immerse oneself in the landscape of the games encourages players to be continually aware of their surroundings, not just from a ludological perspective, but from a narratological one too. These despairing environments do far more than just provide the player with an oppressive tension, they actively convey narrative and context to the surrounding environment. As is commonplace in the *Souls* series, the landscape is not just beautiful, but informative: if you can see it, then most likely, you will explore it at some point (and through exploration, the player can better understand the narrative of *Dark Souls*). In locations such as Anor Londo, Ariandel, and Lothric Castle, we can see clear signs of an unending struggle against the 'dark soul of man'. Mountains of corpses strewn across ramparts, Corvians (crow-like humanoid creatures that dwell in the painted realms) senselessly slaying their own kind, and knights loyal to their gods (even in death) teach the player about the history of the world they are in. In Anor Londo, the player will note the clear erasure from history of Gwyn's first-born son (The Nameless King). Monuments erected to the royal family become disjointed where statues to the first-born son have been either destroyed or removed. Through observing this landscape, the player becomes aware that, at some point, Gwyn disowned his firstborn, though, until *Dark Souls III*, we are unable to determine why. Narrative is everywhere in this game, and the world itself is no exception. As we observe the end of the world in *Dark Souls III*, we see a convergence; the world collapses in on itself as if preparing to start the cycle anew. Ever present in this view is the eclipsed sun, mirroring the accursed Darksign (a symbol of the undead). From this image, we can observe how the cycle is coming to an end. The age of fire is finally coming to an end, and the dark age of man dawns. The end of the *Souls* series provides the player with a sense of resolution in that, this seemingly eternal cycle of linking the first flame is finally closing. With the usurping of the flame, and the new painted world, with the dark soul as its pigment, Lordran is thus allowed to die, so that it may be reborn. This narrative of a dying world and its dying inhabitants posits death as a liberator of the enslaved and corrupted. This world that has been torn apart time and time again by war and darkness, is finally granted the opportunity to rise from the ashes like a phoenix, and players thus see death as necessary for progress. In repositioning the player's attitude to death, it once again shifts the player's understanding of difficulty, and what it is to fail in this game – to die is no longer to fail. Sometimes, it can be the catalyst for success (Penfold-Mounce, 2018).

In examining death and undeath in this narrative, we cannot ignore the role of the Darksign, and the undead plague. The curse of the Darksign was placed upon humanity in an attempt to stifle the lineage of the Furtive Pygmy, and maintain Lord Gwyn's Age of Fire. In doing so, Gwyn cursed humanity with a deteriorating sense of self, and an immortality that fed upon it. Everyone slowly, but surely, lost their minds, memories, and even themselves – ultimately becoming hollows (immortal, soulless, and purposeless husks of creatures afflicted by the curse of the Darksign). The player is encouraged to understand the curse of the undead, embrace it, but not to be consumed by it, and is told that:

> Once, the Lord of Light banished Dark, and all that stemmed from humanity. And men assumed a fleeting form. These are the roots of our world. Men are props on the stage of life, and no matter how tender, how exquisite .... A lie will remain a lie. Young Hollow, knowing this, do you still desire peace? [...] Peace grants men the illusion of life. Shackled by falsehoods, they yearn for love, unaware of its grand illusion. Until, the curse touches their flesh. We are bound by this yoke. As true as the Dark that churns within men. (*Dark Souls II*)

Here, Aldia explains that Gwyn placed the curse upon man, in an effort to create the 'illusion of life', and (falsely) project the fallacy of the continued age of fire. In *Dark Souls III*, the 'Repair' sorcery explains that 'light is time, and the reversal of its effects is a forbidden art'. Armed with this information, we can therefore understand that darkness creates an absence of time, and thus Gwyn, ironically, appears to banish the stagnation of time. As the first flame begins to sputter and die, so too does the progression of time. Peace in this instance refers to the linking of the flame, the metaphorical casting of another coal into the pit – merely postponing the inevitable. Aldia questions the undead's desire to link the flame – the illusion of life – a lie that will only hold up for so long. This notion is further evident when the player interacts with the sleeping Filianore and destroys the egg that appears to keep her asleep. In this moment, the player sees a burst of light that hurtles them to the end of days – where only a few people remain (namely Slave Knight Gael). It is through this time-warp that the player is able to bring about change – killing Gael, and acquiring the Dark Soul to be used as pigment for the new world. When considering this continued effort to stagnate the age of fire, and the creation of undeath in an attempt to do so, *Dark Souls* posits death and undeath as a means to an end (that end being Gwyn's desire for prosperity). This resembles a Lacanian sense of desire in that 'meaning [presents once more,] a void reference in trying to cover over a hole that contains would-be secrets about who and what we are' (Ragland, 1993, 100). In this sense, we can understand meaning as the act of exiling humanity in an effort to cover up the greed, hostility, and delusions of grandeur shown by the ruling classes of *Dark Souls*. As such, the player is encouraged to challenge their understanding of this world, and encouraged not to believe the surface value of it – that the truth lies in a far deeper, and far darker reality. Rather than presenting a linear story for the player to experience, *Dark Souls* actively encourages the player to make their own decisions and interpretations, and to understand the game through what they have observed.

This new world, painted by a nameless young girl (known by fans as Aria), stands to break the cycle of the old worlds, and thus, as the plot of the games develops, we are, once again, thrust towards concepts of rebirth and re-envisioned futures with the resurgence of the painted worlds; in this instance, Ariandel. In the 'Ashes of Ariandel' DLC, we are told of how new worlds are created, how, 'when the world rots, we set it afire. For the sake of the next world. It's the one thing we do right, unlike those fools on the outside' (Corvian Settler, Ariandel,

Dark Souls III). In the painted worlds, the inhabitants reside in relative peace, then, when the rot of the outside world seeps in and begins to destroy their world, they flee into a new one, and the cycle repeats. However, there appear to be rules in relation to this process, as, 'those who aren't ken to fire cannot paint a world. Those absorbed by fire, must not paint a world' (*Dark Souls III*). This dialogue throws into question the role of the linking of the flame in the destruction of the painted worlds, as only those with knowledge of fire (thus, ken to it), can do so. Aria must see fire in order to be inspired for her painting, and that, by using the Dark Soul as pigment, her world will become a 'cold, dark, and very gentle place. So that it might make a home for someone, someday' (*Dark Souls III*). Aria's dialogue here challenges the purpose of the undead, and the presupposed successes of the previous games. Where before the player may have felt a sense of completion in linking the flames, they are now encouraged to believe that this was, in fact, the wrong decision. This narrative of cyclical rebirth thus reinforces the concept of death as cathartic release from a stagnant infinitude. The players' previous victories are now their greatest failures to the inhabitants of this world, and a sense of abject realism begins to consume the player.

When analysing the interactive world of *Dark Souls,* one cannot ignore the narrative of Gwyn, Lord of Cinder. As the final boss of the first game, players initially found this moment of the game to be extraordinarily underwhelming; especially in comparison to some of the game's previous boss fights. However, at this point, Gwyn is hollowed; a husk of his former self, and no longer the powerful lord that he once was. His movements are staggered, and reckless. Unbecoming of a god, they reflect the countless other hollows that roam Lordran, and reveal to the player the reality of what has transpired in the kiln of the first flame. In linking the flame, with the hopes of continuing the age of fire and stifling the abyss, Gwyn has sacrificed all that he is. Cursed, and driven mad as his soul is torn from his body by the first flame, Gwyn becomes an immortal hollow; an unthinking, undying wretch who wanders aimlessly, and attacks anything in sight. By fragmenting cryptic narratives, *Dark Souls* rewards players for their intuition with lore. The more the player engages with the game, the more they learn about the world in which they find themselves. In shifting the onus placed on the player to engage with the narrative from an optional aspect of the game, to the only way in which the game makes total sense, FromSoftware ensures that players are actively seeking out the details of the non-linear storyline.

With this in mind, the player should feel conflicted at slaying not only a god, but also a hero, and a tyrant. Gwyn led the assault on the everlasting dragons; his 'mighty bolts peeled apart [the] stone scales' that granted them immortality (*Dark Souls*). Gwyn forged the Age of Fire and led Lordran in prosperity. However, as expanded upon later, Gwyn tried to stifle the rise of humanity and the abyss; trapping the descendants of the furtive pygmy in the 'Ringed City' where he hoped that their curse would consume and eradicate them. He placed his daughter, the aforementioned Filianore (VaatiVidya, 2017b), in the city, giving its inhabitants the task of guarding her: a sacred task bestowed upon them by a god, but in reality, it was a means of ensuring that they remained with a sense of purpose. In case that was not enough, Gwyn posted demons and even a dragon to the city, in the hopes of deterring/dispatching any potential visitors or escapees. There is a clear undertone of racial and class disparity; a supremacy felt by the ruling class born out of fear, for humanity fought alongside the gods, and their powerful knights returned, empowered, with the heads of everlasting dragons (VaatiVidya, 2017b). In creating an underlying race and class narrative, *Dark Souls* blurs the line between tyrant and hero. In your fight against Gwyn, you are not only killing the god who brought about prosperity, and freedom from the oppressive rule of the everlasting dragons, but you are slaying a tyrant, who cursed humanity for the sake of his own people. An equally oppressive figure that enslaved, entrapped, and slaughtered the innocent out of blind fear. Upon completing the first game, you either link the first flame, and allow the age of fire to stagnate and continue further, or you emerge as a lord of darkness; greeted by the primordial serpents who manipulated you for their own gain. The player is given the illusion of choice in a dystopic and apocalyptic world.

Furthermore, we can better understand this narrative moment when we analyse the music sequence played throughout, and observe the audio-narratives within the game. In Motoi Sakuraba's 'Gwyn, Lord of Cinder', we experience an almost infinite sense of melancholy, confusion, and irresolution. The song explores Lydian and Mixolydian modes of C Major from the perspective of F, progressing to F Lydian and G Mixolydian – shifting to E Phrygian and D Dorian (interspersed with A Aeolian and B Locrian at the end), before (finally) circling back around – purposefully avoiding Ionian so that the listener has no sense of resolution. The opening melody creates an immediate sense of tension and unease with its beautiful, yet jarring, piano that quickly fades in from a gentle hum to a hammering ring. As it progresses, it builds pace while maintaining an unnerving sense of repetition in an active effort to increase tension, drawing focus to the tumultuous world in which the players find themselves. This repetition mirrors the cyclical nature of time and order in the world of Dark Souls, and its continued efforts to rouse the player from any sense of comfort. The final part of the song utilises polychords as it intentionally denies the player any sense of musical resolution – thus further increasing the overarching sense of melancholia, and generating a Gothic vision of the sublime in this audio-narrative. The cyclical nature of the song is intended, not only as a means of creating continuous music throughout the battle, but also as a method of emphasising the clear lack of resolution in the game's narrative overall.

Upon completing the game, the world is not a better place – almost nothing changes, and the cycle is allowed to carry out once more. This morose atmosphere is intended to jar the player as much as the fight itself, and disillusion them to the notion of resolution. While it is commonplace that 'the [music] rhythmically and/or melodically mimics the on-screen action', in this particular moment, the music mimics the narratological emotion, rather than the on-screen action; a reoccurring trope in the *Souls* series (Grimshaw, 2014, 119). By presenting the player with this melancholic audio-narrative, players face an undertone of confliction, and immediately find it an uncanny and intense experience (rather than the presupposed grandeur and antagonistic qualities usually attributed to 'final boss' music). Once again, FromSoftware seeks to challenge our conventional experience of ludo-Gothic narrative experientiality, and encourages us to reconsider both the value of death and its role within contemporary Gothic narrative. This final moment in the game sees the player dethrone a god in an effort to save the world; a god who, in their own right, did the same thing. This narrative instance is less about scaring the player, inasmuch as it is an attempt to displace them. The constant theme throughout the series is that, the perpetuated plotline of 'defeat the bad thing/escape the bad thing, and everything will become utopian once again', is ruthlessly cast aside in favour of a darker, less satisfying ending: that this is a dying world, and there is no 'real' escape. There is no catharsis for the player, only a pained sigh of relief as they prepare to begin the cycle once more in New Game+ (wherein the difficulty is further increased, and the player may bear witness to new enemy spawn locations and items – an arguably underwhelming reward). Throughout the game, players are constantly reminded of the true terror of this world – not just through the written words and mechanics the player will experience, but through the more sensorial aspects of storytelling. The subtle (yet effective) audio-narratives in the *Souls* series contribute to a much larger, and more lasting sense of dread in this way.

Conclusion

Examining the role of narrative engagement reveals to us how contemporary horror ludo-narratives are changing. *Dark Souls* heralded a shift from the false dichotomy of 'survival horror' and 'surviving horror', to the broader spectrum of video-ludic gothic experientiality. As we have seen in the scholarly works of Penfold-Mounce, and in the cultural works of VaatiVidya and others: in this new hybrid genre, Gothic experience and narrative are not limited to the notions of 'escaping the bad thing' or 'defeating the bad thing' they exist outside of this fallacy – presenting a world in which you can neither escape or defeat the overarching horror of a non-linear narrative – *Dark Souls*. As discussed throughout this chapter, in this series, the player is tasked with not only earning the fragmented nested narrative, but also piecing it together themselves, and interpreting it as they see fit. This narrative methodology, and the onus placed upon the player to interact with, and immerse themselves in the lore of the games, has led to a greater player investment, as seen with the popular cultural works of YouTube content creators. Alongside this, FromSoftware has generated a significant volume of populist scholarship around their key titles (as discussed with Mecheri and Romlieu, 2017, 2018). This cultural phenomenon of hybrid ludo-narrative immersion is aimed at closely analysing the narrative and mechanics of the games so as to better understand the Gothic ludo-narrative. In a game where the player is rewarded with plot, and broken down by an oppressive environment, personal growth is inevitable. Hardcore fans of the *Souls* series became inherently proud of their ability to overcome such unapologetic difficulty, relishing in no longer being led by the hand through a thinly veiled story. Instead, we are thrown headfirst into chaos. We engage with the denial and stagnation of a dying age so as to prevent the horrors of an unknown, human, future. A future wherein, the gods are dead, faith and humanity are but a memory, and only darkness remains. *Dark Souls*' cyclical, non-linear timeline actively seeks to disorient the player, and continually challenge their understanding of the dying realm of Lordran. In doing so, we as players are forced to re-evaluate our understanding of, and approach to, death in the Gothic ludo-narrative. Throughout the series, death is repositioned as liberating, a necessary component for progression (both personally and socially). In some cases, as with the initial encounter with Seeth the Scaleless (the aforementioned mortal abomination), death is vital to the progression of the game and narrative. Death is therefore no longer the failure that it used to be, but rather, a step in the realm of personal development. In this new hybrid Gothic narrative mode, we see a shift in linear narratives spoon-fed to a bored audience – narrative experience is now earned, and FromSoftware has established a transnational, intermedial ludo-Gothic experience.

Interactions and Sensoria

Bandai Namco. 2016. *Dark Souls: The Art of the Trilogy*, Europe, Bandai Namco.

EpicNameBro. 2015. Let's play dark souls playlist, *YouTube*. Available at: https://youtu.be/KUqg8XtAHnc

Tinwell, A., Grimshaw, M. and Williams, A. 2010. Uncanny behaviour in survival horror games, *Journal of Gaming and Virtual Worlds*, 2(1), 3–25.

VaatiVidya. 2016–2019. Dark souls 3 lore playlist, *YouTube*. Available at: https://youtu.be/fs5fNUKn08c

VaatiVidya. 2019. The joint timeline of soulsborne, *YouTube*. Available at: https://youtu.be/vT_4jf02wsU

References

Black, H. 2016. Why Blighttown Really Matters (Dark Souls) – Writing on Games, *YouTube*, 08 October. Available at: http://bit.ly/33eZxzu

Coward, M. 2018. Survival Horror and Surviving Horror: Wolfenstein, Traumatic Play and the Social Gothic. In *Proceedings of Gaming the Gothic*, University of Sheffield, 13 April.

Elsaesser, T. 2009. The Mind-Game Film. In *Puzzle Films – Complex Storytelling in Contemporary Cinema*, Ed. W. Buckland, pp. 13–42, Malden, MA, Wiley Blackwell.

Frasca, G. 2003. Ludologists love stories, too: notes from a debate that never took place. In *DiGRA: Digital Games Research Conference Proceedings,* Utrecht University, 92–99.

FromSoftware. 2011. *Dark Souls*, Bandai Namco.

FromSoftware. 2014. *Dark Souls II*, Bandai Namco.

FromSoftware. 2016. *Dark Souls III*, Bandai Namco.

FromSoftware. 2018. *Dark Souls Remastered*, Bandai Namco.

FromSoftware. 2009. *Demon's Souls*, Bandai Namco.

Grimshaw, M. 2014. Sound. In *The Routledge Companion to Videogame Studies*, Eds M.J.P. Wolf and B. Perron, pp. 117–124, London, Routledge.

Hunter, P.L., Bowman, N. and Banks, J. 2015. The Enjoyment of Griefing in Online Games, *Journal of Gaming & Virtual Worlds*, 7(3), 243–258.

Jenkins, H. 2006. *Convergence Culture: Where Old and New Media Collide*, New York, NY, New York University Press.

Johnson, M. and Woodcock, J. 2017. 'It's Like the Gold Rush': the Lives and Careers of Professional Video Game Streamers on Twitch.tv, *Information, Communication & Society*, 22(3), 336–351.

Kaytoue, M., Silva, A. and Raïssi, C. 2012. Watch me Playing, I am a Professional: a First Study on Video Game Live Streaming. In *WWW '12 Companion: Proceedings of the 21st International Conference on World Wide Web*, pp. 1181–1188. New York, NY, ACM.

Kirkland, E. 2009. Narrative in survival horror videogames. In *Horror Video Games*, Ed. B. Perron, pp. 62–78, Jefferson, McFarland & Company.

Kokonis, M. 2014. Intermediality Between Games and Fiction: the 'Ludology vs. Narratology' debate in Computer Game Studies: a Response to Gonzalo Frasca, *Film and Media Studies*, 9, 171–188.

Kroener, O. 2014. Breaking narrative: Narrative: Narrative Complexity in Contemporary Television. In *Critical Reflections on Audience and Narrativity: New Connections, New Perspectives*, Eds V. Marinescu, S. Branea and B. Mitu, pp. 77–88, Stuttgart, Ibidem.

Krzywinska, T. and King, G. 2006. *Tomb Raiders and Space Invaders: Videogame Forms and Contexts*, London: I. B. Tauris.

McManus, A. and Feinstein, A.H. 2006. Narratology and Ludology: Competing Paradigms or Complementary Theories in Simulation, *Developments in Business Simulation and Experiential Learning*, 33, 363–372.

McWhertor, M. 2012. Dark Souls' Most Mysterious In-Game Item was a Prank, Says Game Director Hidetaka Miyazaki. *Polygon*, 5 November. Available at: http://bit.ly/2puu4uF

Mecheri, D. and Romieu, S. (Eds) 2017. *Dark Souls Beyond the Grave. Volume 1: Demon's Souls – Dark Souls – Dark Souls II*. Toulouse, Third Editions.

Mecheri, D. and Romieu, S. (Eds) 2018. *Dark Souls Beyond the Grave. Volume 2: Bloodborne – Dark Souls III*. Toulouse: Third Editions.

Murray, J.H. 2005. The Last Word on Ludology v Narratology in Game Studies. In *DiGRA: Digital Games Research Conference,* Vancouver, Canada, 17 June.

Penfold-Mounce, R. 2018. *Death, the Dead and Popular Culture*, Bingley, Emerald.

Perron, B. and Wolf, M.J.P. (Eds) 2003. *The Video Game Theory Reader*, London, Routledge.

Picard, M. 2009. Haunting Backgrounds: Transnationality and Intermediality in Japanese Survival Horror Video Games. In *Horror Video Games*, Ed. B. Perron, pp. 95–120, Jefferson, McFarland & Company.

Ragland, E. 1993. Lacan, the Death Drive, and the Dream of the Burning Child. In *Death and Representation*, Eds S.W. Goodwin and E. Bronfen, pp. 80–102, London, Johns Hopkins University Press.

Shinya, Y. 2004. Funeral Rites and Changing Perceptions of Death in Contemporary Japan, *Mortality*, 9(1), 27–41.

Taylor, L.N. 2019. Gothic Bloodlines in Survival Horror Gaming. In *Horror Video Games*, Ed. B. Perron, pp. 46–61, Jefferson, McFarland & Company.

VaatiVidya. 2017a. Dark souls 3 ▶ Angel's Egg & Filianore, *YouTube*, 02 June. Available at: http://bit.ly/2WJIFyu

VaatiVidya. 2017b. Dark souls 3 ▶ Story of the Ringed City [Part 1], *YouTube*, 03 May. Available at: http://bit.ly/34pYwES

Chapter 7

'He Died a Lot': Gothic Gameplay in *What Remains of Edith Finch*

Ewan Kirkland

Abstract

This chapter explores how the narrative-based walking simulator *What Remains of Edith Finch* ludifies traditions of Gothic fiction. Combining Gothic themes of death, the family and the family curse, the game involves the protagonist investigating her abandoned childhood home where every family member died a dramatic and untimely death. Sealed rooms, preserved since their inhabitants' demise, contain shrine-like displays including a document of some form allowing players to experience the last moments of each Finch. Play involves penetrating these spaces, according to the ludo-Gothic emphasis on boundary crossing, piecing together interactive narrative fragments consistent with Gothic fiction's patchwork storytelling. In accessing each lost manuscript, players engage in a generically specific process of multi-media trans-subjectivity, experiencing various first person perspectives and engaging with numerous gameplay interfaces. The title's series of ambiguous unreliable narratives, its refusal of a consistent subjective position, and unreal dream-like sensation contribute to the game's Gothic atmosphere. In a restriction of videogame agency and control, consistent with horror games, no player option is available other than to complete each pre-determined death. Gothic pastiche, a compulsion to repeat the past, and the embalming processes of photographic media are variously employed across these sequences. Play evokes the melancholy heroine, consumed by maternal loss, masochistically replaying her family's sorrowful past, hunting for lost objects and exhuming the ghosts of her history. With its nested narrative, morbid preoccupation and ambivalent supernatural presence, the game effectively translates Gothic traditions into the videogame medium.

Keywords: Family; Gothic; home; death; *what remains of Edith Finch*; melancholy

Death, Culture & Leisure: Playing Dead, 95–107

doi:10.1108/978-1-83909-037-020201014

Introduction

This chapter is an exploration of Giant Sparrow's acclaimed 'walking simulator' *What Remains of Edith Finch* (2017), drawing on a combination of videogame scholarship and Gothic studies. In an appropriate reflection upon the game's title, Shane Snyder considers *Edith Finch*, alongside *Dear Esther* (The Chinese Room, 2012) and *Everybody's Gone to the Rapture* (The Chinese Room and SCE Santa Monica Studio, 2015), to be a videogame unfolding within an abandoned space, involving 'an exploration of what has been left behind' (Snyder, 2018, 9–10). The game tells of a young woman's return to her deserted family home following the death of her mother. Narrated by the eponymous protagonist, *Edith Finch* is structured as a journal, a testimony and an ambiguous warning left by its central character to her unborn child, detailing the family curse which has claimed the lives of each Finch before her. Sealed rooms throughout the house contain shrine-like spaces displaying paintings of family members, objects reflecting their lives, and a document of some description. Engaging with this text produces a brief interactive sequence, teleologically structured to end in the demise of its respective family member. Play involves investigating the dynastic mansion, learning how generations of Finches were variously crushed, drowned or decapitated. This account follows Edith's path through the house, examining the twisted chronology she encounters, illustrating how the game reproduces themes of Gothic fiction while translating literary tropes and traditions according to the conventions and affordances of the videogame medium. The Gothic is a cycle preoccupied with death and deaths, families and family curses, the investigation of uncanny ancestral homes and unclear ancestral histories. *Edith Finch* synthesises these themes in an interactive experience which effectively conscripts the player as an ambivalent presence, knowingly initiating the fateful sequence of events which leads to the death of each Finch family member.

Gaming the Gothic

'Ultimately', Charles L. Crow (2009, 15) concludes, 'all Gothic stories are family stories', and the structural centrality of the family tree to *Edith Finch* roots the game in this Gothic framework. As play progresses the branches of the protagonist's notebook fill with successive images of Finch generations. Continuing, or more likely establishing this tradition, Edith's great grandmother Edie senior painted family members' portraits on wood cross sections, commemorating their birth and death dates, ritualistically placed in their room following their demise. Clive Bloom (2010, 64) identifies death as an obsession of the Gothic, and death pervades *Edith Finch*, from poisonous foxgloves surrounding the family estate, to empty bird cages in Edie senior's room, to taxidermy pets and hunting trophies, to the house's back yard cemetery. The game's title itself plays on various morbid dualities. 'Remains' refers to both what is left behind and the dead body itself, while the moniker might signify either Edie senior or Edith junior, the nonagenarian's sprawling mansion or the young woman's notebook. Twin Gothic preoccupations of death and the family intersect in the dynastic curse, a theme evident in the earliest Gothic novel (Punter and Byron, 2010, 29). The supernatural enjoys an ambivalent presence in *Edith Finch*, and Gothic media, and the Finch family home is undoubtedly haunted by the remains of its departed inhabitancy. The past saturates this game, in the nostalgic, regretful tone of its narrator, in the objects arranged in the rooms of each dead inhabitant, and in the architectural organisation of the building, testimony to the Finch family's peculiar form of mourning.

Complementing such Gothic preoccupations is a gameplay structure which involves a more ludic translation of generic tropes. Within Gothic literary criticism, Eve Kosofsky Sedgwick (1986, 12) identifies a 'spatial model' wherein the self is 'massively blocked off from something to which it ought normally to have access'. This 'something' may take various forms, including 'its own past' and 'the details of its family history' (Sedgwick, 1986, 12). Such a description neatly fits the circumstances of *Edith Finch*'s protagonist. Having entered the house, a chore in itself, Edith encounters a succession of sealed doors, crudely gummed shut. Peepholes offer tantalising glimpses into the elaborate spaces within. Exploring proximities between horror fiction and videogames, Laurie N. Taylor (2009) coins the term 'ludo-Gothic' to describe the transformation of literary tropes into gameplay, echoing Sedgwick's 'spatial model' in emphasising the theme of boundary crossing within Gothic videogames. The structuring principle of *Edith Finch* involves gaining progressive access to those rooms and the strange histories they contain. Edith's story is pieced together, patchwork style, through a series of discovered manuscripts, another device of Gothic fiction (Mighall, 2007, 113), each text activating an interactive micronarrative which culminates in the death of a Finch family member. Sedgwick (1986, 9) considers similar 'tales within tales, changes of narrators, and such framing devices as found manuscripts or interpolated histories' characteristic of the Gothic novel. Consistent with Gothic literature's tendency to combine different genres and formats, these 'nested' stories involve a range of media, including a diary, a comic book, a letter and an animated flick book. Exploring videogames' incorporation of different media, such as photography, video, tapes and databases, Christian McCrea (2009, 222–223) draws attention to the observed relationship between modern media and representation of the dead. Blending traditions of Gothic fiction and survival horror, the documents encountered throughout *Edith Finch* explicitly facilitate a dislocation from time and space which affords both protagonist and player intimate contact with Edith's long departed relatives.

Complementing Sedgwick's (1986) 'interpolated histories', Sue Morris (2002, 93) draws on Marxist philosopher Louis Althusser, employing the framework of 'interpellation' to describe digital games' use of the first person perspective to effectively suture players into particular subjective positions. Both terms seem appropriate descriptions of the process whereby dead Finches are reanimated through participants assuming the protagonists' role within their stories. The player of *Edith Finch* is both inserted into the story of the unlucky family and made to assume the identity of each doomed member. A highly reflexive game, *Edith Finch* thereby plays with different subjectivities in a manner consistent with Gothic media, unsettling traditional boundaries between self and other, challenging concepts of a fixed and unified identity, and allowing players to experience different physical and psychological conditions. The first story players find tells of Molly Finch, a young girl who apparently died while locked in her room without supper. The consumption-themed dream which unfolds sees the player, as the reader of Edith's journal, as Edith reading Molly's journal, as Molly dreaming, transform from cat to owl to shark to sea monster. Across this sequence the player crawls, leaps, flies, swims and slithers from the optical perspective of each animal, consuming a bird, rabbits, seal and eventually human beings along the way. The final section involves directing an eel-like creature up a sewage pipe into Molly's room. It hides under her bed as the girl writes of soon being eaten alive while she sleeps.

There is an ambiguous complexity to this consuming force, characteristic of the ambivalent monsters of Gothic fiction. The sea creature is at once an imaginary projection of the character Molly, a surrogate for the player directing the beast into the room of the sleeping child in order to complete the sequence, and a symbolic representation of the deadly curse which haunts the Finches. Other subjective positions offered across the game include a man sealed in an underground bunker, a father and daughter on a hunting trip, and a boy flying a kite, all played from the first person perspective. Adapting psychoanalytic film criticism, with some qualification, Morris (2002, 89) draws parallels between videogame play and the immersive dream-like state of cinema spectatorship. Such a position seems appropriate for Molly's fevered fantasy which establishes the game's phantasmagorical tone early in the experience. This unreal quality, consistent with the tone of much Gothic fiction, persists in the more fanciful sequences which narrate each Finch member's death, including a baby with apparent telekinetic powers, a young artist disappearing into his own creation, and a teenage boy suffering from hallucinations.

Gothic Interpellation

Discussing militaristic first person shooters, Morris (2002, 89) argues one significant compromise to the comparative dream-like qualities of cinema is the level of control players exercise over on-screen action. Despite adopting a similar interface, *Edith Finch* is far removed from the traditional shooters Morris considers. The player has no weaponry, confronts no adversaries, and has minimal impact upon their environment. Beyond walking, slowly, around the Finch estate, interactions are heavily prescribed and limited to environmentally specific actions such as opening a door, turning the key in a music box, investigating a document in a dead relative's bedroom. Such limitation of player agency is central to Tanya Krzywinska's (2002) discussion of horror videogames, in a chapter which generically situates the ludic structures informing *Edith Finch*'s mediation on history, death, fate and the family curse narrative. Across horror games, Krzywinska argues, oscillation between player action and inaction, agency and its withdrawal, maps onto horror themes whereby 'supernatural forces act on, and regularly threaten, the sphere of human agency'. A 'moral occult', analogous to the infrastructure of the game's coded design works as a 'predetermined transcendent force', leading the player through the ludic labyrinth, variously restricting and facilitating progression, often producing 'an experience of being subject to a predetermined, extrinsic, and thereby, Othered force' (Krzywinska, 2002, 207–208).

Much of *Edith Finch*, from its past tense temporality to the foreshadowing clues scattered across the family home, enhances this sense of pre-determination. Fate controls Edith's sequential investigation of the house, and more profoundly, the death scenes players replay at regular intervals. Here, interaction is extremely limited. Molly's story, the first players encounter, is one of the more elaborate, offering many actions to perform, bodies to inhabit, and spaces to explore beyond the cluttered Finch estate. The subsequent story of Odin Finch takes the form of a non-participatory viewing slide. Reproducing a common trope that David Punter (2014, 24) observes in American Gothic fiction, it tells of Edith's descendants' departure from Europe, intent on escaping the family curse, only to bring it with them. The Scandinavian house on which the eccentric Odin sailed across the Atlantic sinks just within reach of the shore, taking the patriarch with it. Calvin's story follows. With characteristic dryness the narrator previously observed, while peering through Calvin's door's peephole, that for seven years Sam shared a bedroom with his dead twin. Their room is a divided space, bisected by a barrier resembling the red ropes used to mark off inaccessible exhibits in a museum. Unclipping this obstacle, players are guided by the narrator's words to climb a low flight of stairs ending in a shrine similar to Molly's, containing a short essay hidden inside a toy space helmet.

Death, duality, divided spaces, secret manuscripts, preserved rooms, shrines and monuments to departed relatives. These are common features across Gothic painting, literature, film and television. Where *Edith Finch* develops, translates and transforms these themes is through the interactive possibilities and limitations of the videogame medium. Players are transported into the position of a boy on a tree swing overlooking a cliff edge. Interactive opportunities are slight. Players can inspect the coastal landscape, the sky, the sea and the other boy stood gazing into the distance. But they cannot leave the swing itself. Experimenting with the analogue sticks results in the child's feet pivoting back and forth, until the swing starts to oscillate in time with this movement. Many indications suggest this sequence will not end well, but the 'pull' established by the game, to employ Jesper Juul's (2010) term for the intuitive input digital games elicit, is to continue swinging Calvin, up and down, higher and higher. Players may even experience the sensation Morris (2002, 87) identifies, unconsciously reproducing the actions of on-screen avatars, rocking in synchronisation with the boy's body. Perhaps the adult player will recall, as Edith confesses to remembering, swinging as a child and being teased by the belief that it was physically possible to rotate a full three hundred and sixty degrees. Eventually, this is precisely what happens. The swing chain wraps round and round the tree branch, until the player, along with Calvin, is flung high into the air and the world fades to white.

All Finch family members are fated to die, are indeed already dead, and the game refuses players any option but to perform their last moments. Interactive affordances are perversely circumscribed to make players complicit in the occulted affliction controlling Finch family history. To this extent, paraphrasing Krzywinska, (2002, 207) players themselves become the ambivalent supernatural force which threatens human life and agency. A painful fit exists between videogame scholar Krzywinska's moral occult and what Gothic literary critic Robert Mighall (2007, 124) calls 'the deterministic structure associated with the theme of the ancestral curse' with which Gothic fiction ambivalently plays. The game also capitalises on the first person interface's ability to position players in the physicality of fictional protagonists. This exploits the medium's capacity, discussed by Jon Dovey and Helen Kennedy (2006, 106), to 're-embody' players through a sense of virtual presence and agency. Molly's story extends the Gothic preoccupation with exploring different, non-human, potentially monstrous, subjectivities. Calvin's more physically grounded scene appears designed to elicit a clearer, consistent, physical relationship between player and playable protagonist. With its intuitive controls and synchronistic movement, the line dividing player and screen is blurred in a manner consistent with the Gothic theme of boundary erosion, dissolution and penetration. In adopting the 'preferred performance' (Dovey and Kennedy, 2006, 115–116) demanded by the game, moving Calvin's legs so they cause his body to rise and fall, players might feel increasingly at one with the boy on the swing. This close proximity between player and Calvin has sinister implications for the nature of the Finch family curse. For the knowing, active, wilful player inhabiting the body of the doomed child, deliberately bringing about the boy's death, suggests a similar knowing, active wilfulness on Calvin's part, almost as though he, like the player, understands they are fated to die, but finds himself helpless to resist the pull.

Compulsions to Repeat

Ambiguity is a recurring feature of Gothic fiction, and Barbara's story is one of the most ambiguous of *Edith Finch*. Characterising a tendency of Gothic fiction towards irony, pastiche and parody, Barbara's interactive flashback assumes a *Tales from the Crypt*-styled comic book story, complete with cel-shaded panels, speech bubbles, sound splashes and hammy commentary provided by a pumpkin-headed host. Its tall tale recreates the quintessential camp fire story, featuring a babysitter and her boyfriend being terrorised by a hook-handed killer, accompanied by the signature tune of John Carpenter's *Halloween* (1978). Such gestures reflexively illustrate the extent death remains a source of fascination for much storytelling, popular culture and media entertainment. Additional self-awareness is provided by Barbara's characterisation as a faded child star. Her bizarre story begins with adult Barbara's invitation to a horror film convention, and ends with the actor surrounded by movie monsters who descend upon her in a peculiarly violent act of adulation. What appears to be a fictional take on the film star's mysterious death gives no clear explanation of what happened to Barbara, how her fate relates to the Finch curse, why her story should be the subject of a commercially available comic book, or how her death relates to the traumatisation of Walter, the brother she was babysitting.

Yet some corroboration of the story is evident in the preserved room where the comic book lies. Barbara evidently surrounded herself with generic Hollywood iconography, merchandise and advertising material promoting her hit monster movie. Throughout her adult life Barbara appears preoccupied with her short-lived career, clinging to relics of her former stardom, living in a mausoleum of memorabilia, even before her room was sealed shut. The debilitating consequences of being trapped in time, fated to perpetuate the mistakes of older generations, is another Gothic theme of the family curse narrative. Eric Savoy (2010, 172) writes of an established tradition in American Gothic whereby children, plagued by the 'sins of the father', find themselves ironically caught in 'an intergenerational compulsion to repeat the past'. Compelled by the player, Edith appears similarly fated to re-enact the final moments of her relatives' lives. Exploratory interludes often entail retracing the movements of previous Finches, following Molly's paw prints, or replaying Barbara's suspenseful descent into the basement. This leads to the story of Walter, in which the theme of repetition is central. Traumatised by whatever killed his sister, and by the forces plaguing his family, the boy hid himself in a bunker built beneath the Finch house. There players experience Walter's monotonous existence, performing the same actions day after day, year after year, until, in an inadvisable moment of optimism, the man emerges from his self-imposed prison, only to be run over by a train.

As Dovey and Kennedy (2006, 116) observe, repetition is uniquely fundamental to videogame play, an experience strongly associated with in-game death. Indeed, gaming often entails watching death sequences played over and over again, until the vanquishing of a particularly tricky adversary is achieved. Reflecting Krzywinska's (2002, 219) discussion of death's ambivalent presence in horror games, dying in *Edith Finch* is not a sign of failure. Quite the contrary. Successful play involves leading consecutive family members to their doom. After Walter's story, Edith follows the mole man's path out into the open. This leads to a brief excursion around the family's clifftop cemetery which Edith notes, with a characteristic mixture of disdain and resignation, the Finches built before the house itself. Here, players are offered the chance to view the original family home, protruding from the sea, marked by a red beacon and clanging bell, as they stand beside a monument to the building sinking, taking Odin Finch with it. Next players enter the room occupied by Calvin's now-adult twin brother following his retirement from the armed forces as a military photographer. The space is filled with medals and trophies, pictures of hunting and military activity, weapons and photographic equipment, stuffed animals and preserved images.

Sam's room, and the photography-based flashback it tells, represents a mediation on the deadly proximities between the camera and the gun. Susan Sontag (1979, 15) famously argued all photographs constitute '*memento mori*'. Photographic portraits cannot avoid ironically reflecting upon their subject's mortality, producing an image which will withstand the ravages of time in contrast to their aging sitter. Furthermore, Sontag observes how popular photography has always enjoyed an intimate relationship within familial structures, being integrated into festivals and rites of passages, even as the institution it memorialised became increasingly under threat from historical forces. 'A family's photograph album is generally about the extended family', Sontag (1979, 8–9) writes, 'and often, is all that remains of it'. Photographs feature throughout the Finch family home. The first players encounter, as if to underline the medium's inherent morbidity, appears on the kitchen fridge beside a notice for Lewis Finch's funeral. Many Finches appear preoccupied with collecting images of their fleeting family, as home videos, hand prints, paintings, and the sketches which progressively fill the family tree of Edith's journal. Connections between shooting guns and shooting photographs, also considered in Sontag's essay, are evident in the subsequent flashback, drawn from still images captured on a hunting trip taken by Sam and Edith's mother, Dawn. Depicted entirely through the camera's lens, this sequence involves the player, as Dawn, photographing various wildlife, and then, as Sam, photographing his daughter's first kill. Reproducing the camera as 'sublimation of the gun' (Sontag, 1979, 14–15), in the sequence's conclusion Sam assumes a position directly behind Dawn's shoulder, taking the photograph the moment she pulls the trigger, capturing the point the bullet strikes its target. Ironically, the final image taken on the trip records how Dawn's wounded kill unexpectedly reared, sending a startled Sam tumbling to his death.

Mourning and Melancholia

The disposition of *Edith Finch*'s titular character parallels that of the archetypal Gothic heroine, described by Bloom (2010, 69), as one defined by melancholia. Sigmund Freud wrote on the qualities of mourning and melancholia, noting their common features. Mourning, defined as 'the reaction to the loss of a loved person' justifiably fits Edith's frame of mind, having recently lost her mother. However, Freud claimed the same circumstances produce in some the 'pathological disposition' of melancholia, characterised by 'a profoundly painful dejection', 'a disturbance of self-regard' and 'a delusional expectation of punishment' (Freud, 1991, 251–252). This might explain Edith's morbid preoccupation with Finch family history, her masochistic repetition of often-fantastical death sequences, and the irrational belief that her very investigation might be reawakening the family curse. A lack of self-regard is inherent in the animalistic position the character adopts towards the house itself. Edith enters through the dog flap, explores the house by crawling through littered passageways, climbs cat-like across the roof, and breaks into the house's sealed rooms as an abject outsider. The discovery that she is pregnant, destined to perpetuate the Finch's troubled family line, combined with her mother's death, might be the cause of Edith's melancholia, and is strangely expressed in the two flashbacks which follow.

The next Finch family death appears through the eyes of infant Gregory, who drowned while unattended in the bathtub. This unsettling story involves players controlling the baby's hand, conducting a collection of bathtub accessories, including a wind-up frog, a flock of toy ducks and a family of plastic dolls, in an enchanted procession around the bathroom. The magical yet disturbingly surreal sequence must conclude with players knocking the tap until water fills the bath. Gregory then transforms into a green-skinned reptile, swimming through reedy waters, before sinking towards the light-filled plug hole. No indication of the baby's fantastical experience is evident in the document lying on Gregory's crib shrine, a retrospective account written by the child's father. Despite the sequence's cheery colours, lively sound effects and classical accompaniment, what players perform might constitute a dark infanticidal fantasy directed towards Edith's unborn child. This is followed by the equally unsubstantiated story of Gus, killed flying a kite while stubbornly refusing to attend his father's wedding. As in Gregory's story, an uncanny omnipotence of thought informs the boy's command of the elements, using the kite to whip up a storm, gathering the furniture surrounding the wedding marquee to violent effect. If Gregory's story reflects Edith's unconsciously murderous disposition towards her unborn baby, this fantasy of a young boy exacting revenge on his distant parent's remarriage might either be an expression of Edith's hostility towards her mother or the absent father of her own child for deserting her. The poem accompanying this story, notably written by Dawn herself, ends on a tone of regretful sadness for leaving her brother alone that day.

Elaborating on Freud, in the context of Gothic fiction, Savoy observes how melancholia typically entails an unresolved relationship with a 'lost object', such as a parent, which the mourner seeks to recover through a process of identification with the thing which they have lost. Consequently, as Savoy (2010, 173) explains, employing appropriately Gothic imagery, 'the ego altered by such identification becomes a kind of unquiet grave that harbours the living dead' effectively brought back to life by the sufferer's efforts to overcome their own internal turmoil. Accordingly, Edith's journey entails exhuming the family home of its ghosts, histories and stories. The sealed Finch mansion is filled with symbolically resonant lost objects, lining the walls as photographs and paintings, populating passageways and wall cavities, and explicitly decorating the shrines commemorating each dead Finch family member. Scholarship interpreting early Gothic fiction as reflecting Western denial and divorcement from death leads Robert Miles (2002, 118) to observe 'a universal spectralization', in which 'language and landscape both find... themselves elegantly suffused with mementoes of the dead'. *Edith Finch*'s environment constitutes an over-determined monument to death's privatisation and domestication, overlaid with the narrator's own posthumous prose. The Gothic heroine's efforts at understanding and reconciliation, triggered by these lost objects, effectively reanimate the dead through identification with their optical perspective. A particularly lost Finch member, Edith's brother Milton does not so much die as disappear. Early in the game, posters featuring the missing child littered the path through the Finch estate, while paint marks many hidden spaces with his own peculiar signature. Milton's sequence is animated through the most primitive of moving media, a flickbook, more charming than uncanny. It shows the artist painting a door, opening it, stepping through, bowing, before closing the door behind him, at which point the animation ends. A typically Gothic self-reflexive mediation on artistry, artifice, animation, and videogame processes of interaction and absorption, Milton's scene whimsically precedes the game's final elaborate death sequence.

While the Gothic family curse narrative originates in supernatural tales set in quasi-medieval locations, Mighall observes how, across the centuries, the literary trope of the ancestral legacy becomes increasingly informed by more scientific understandings. One strand blames aristocratic hereditary transmission for the perpetuation of illnesses across the generations (Mighall, 2007, 90). Another trope emphasises the debilitating consequences for those unable to overcome unpleasant memories from the past (Mighall, 2007, 106). Throughout *Edith Finch* the eponymous narrator frequently considers their family's morbid preoccupation with death to be the cause of their malaise. Depression is certainly implicated in the suicide of Edith's older brother Lewis, whose story is one of traumatised withdrawal from reality. In recreating Lewis' death, players enact the young man's repetitive job, guillotining fish heads at the local cannery, while simultaneously guiding an imaginary monarch around a progressively expansive three dimensional space. Lewis, and the player, becomes increasingly absorbed in this hallucinatory videogame-like existence until it fills the entire screen. The sequence concludes in an out-of-body first person experience, where players observe Lewis mechanically working at the cannery. The teenager remains at his station, surrounded by blood, guts and dead fish, while his imaginary self ascends a conveyor belt towards a floodlit window, opening onto a church-like palace thronging with adoring crowds. But the sense of elation is short lived. Back in the real world a half-completed portrait of Lewis stands at Edie Senior's easel. The young man's funeral announcement hangs in the Finch family kitchen. The structure and conclusion of every flashback, and the game's requirements of the player, have been clearly established. As the narrator, Lewis' psychiatrist, says, 'there was only one thing left to do'. The player leads King Lewis up the aisle and places his head in the guillotine necessary for his coronation ....

Going Home

As argued throughout this chapter, *What Remains of Edith Finch* draws upon multiple Gothic themes, tropes and traditions. These include the layering of nested stories and multiple voices, the sealed preserved room, the hidden manuscript, the penetration and erosion of boundaries, a sense of fated pre-determination, melancholy and mourning, and the ambivalent family curse. These elements are present across the stories *Edith Finch* tells, while also being translated into gameplay mechanics and the ludic activities participants perform. As players breach the secret entrances to various locked spaces, they become an active participant in unearthing the forbidden knowledge and hidden histories which preoccupy so many characters of Gothic fiction. Through frequent shifts of player identification, optical perspective and user interface, the game refuses any stable, subjective, consistent or coherent position from which to assess the Finches' fantastical saga. By forcing an enactment of the last moments experienced by each fated family member, players become complicit in the family curse, which retains a typically Gothic ambivalence throughout. *Edith Finch* is also filled with Gothic images of death and metaphors for dying. Flashback sequences end with characters ascending into the air, stepping through a door, moving towards a bright light, going to sleep and going home. Symbols of death pervade the game, the first of which, attentive players might recall, is the cluster of lilies held in the hand of the first character whose optical perspective they inhabit. This is the reader of Edith's journal, gradually revealed to be her then unborn child, on a ferry journey to visit his mother's grave.

Following Lewis' death, Edith comes to the final resting place of her journey. This involves climbing the staircase to her childhood bedroom, another dream-like space, decorated with fluorescent stars, childhood toys and origami birds. Sadly resonant mementoes reflecting previous death scenes are placed across the room, including the frog from Gregory's bath, Sam's deer, figures resembling King Lewis and his subjects. Edith lies on the bed, her distended belly clearly showing, takes a neo-medieval quill pen from a case, and begins writing her own story. This tells of her last night in the house, when she found a secret entrance into the family library, the only sealed room until this point inaccessible. The library often holds portentous significance in Gothic fiction, a genre frequently reflecting upon processes of writing, storytelling and the recording of history. Here, the young Edith finds a hand-written account, penned by great grandma Edie, of her own return home. This home is the house, sunk at sea, some distance from the shore of the Finch estate, which Edie claims to have visited the night her great granddaughter was born. An ocean earthquake caused the tide to shrink, allowing the woman to cross the sea bed in a hallucinatory sequence, resonant with psychoanalytic aspects and imagery. Edie recalls encountering a procession of lost, forgotten, yet familiar things along the way, until she reached the building which emerges from the mist like a Gothic mirage. The surreal sequence combines recurring themes of loss, nostalgia, memory, and return to the abandoned family home. Edie's movement as she opens the gate to the sunken mansion recalls that of Edith, at the beginning of the game, entering the family grounds. Only whatever she discovered in that house appears long lost. Abruptly Dawn interrupts her daughter's reading in the forbidden library, tears the manuscript from Edith's hands, and together they leave the home along with everything it contains, including the elderly matriarch.

Three brief sequences follow, ending with the player placing Edith's hand in Dawn's the moment her mother dies of an unspecified illness. *Edith Finch*'s final interactive scene is a perverse combination of morbidity and maternal loss, characteristic of overlapping concerns within Freud psychoanalysis, Gothic criticism and the game this chapter explores. Players participate in the delivery of the latest Finch child, repeating the curse's cycle of birth and death, in a traumatic sequence implicitly resulting in the demise of the game's protagonist. From first person perspective, players guide Edith's baby through his mother's birth canal, towards the light of the delivery room. Given the nesting order of narratives across the game, from the manuscripts located on Finch family shrines, to the recounting of their stories in Edith's journal, to the imagining of her experiences from the perspective of the child reading this account, primacy is given to the last remaining Finch who is effectively the author of everything the player plays. The sense of melancholia which pervades the title may not be that of Edith for Dawn. Instead it might be Edith's child mourning the mother he never met, whose birth led to her death, whose legacy is restricted to a journal documenting their fantastical family home and history. The final shot sees the child, all that now exists of the

Finch lineage, closing the journal detailing Edith's story, while stood by the grave containing his mother's earthly remains, as the scene rises upwards to reveal the Finch family mansion standing defiantly on the cliff face. It is a suitably Gothic end to a highly Gothic game.

Interactions and Sensoria

Burton, T. (Director). 2012. *Dark Shadows*, Warner Bros. Pictures and Roadshow Entertainment (113 minutes).

Capcom. 2005. *Haunting Ground*, Capcom.

Infinite Fall. 2017. *Night in the Woods*, Finji.

Jackson, S. 1962. *We Have Always Lived in the Castle*, New York, NY, Viking Press.

King, S. 1974. *Carrie*. New York, NY, Doubleday.

Kirkland, E. 2021. *Videogames and the Gothic*, London, Routledge.

Konami Computer Entertainment Tokyo, 2001. *Silent Hill 2*, Konami.

The Fullbright Company and Bitworks. 2013. *Gone Home*, The Fullbright Company, Majesco Entertainment, and Annapurna Interactive.

References

Bloom, C. 2010. *Gothic Histories: The Taste for Terror, 1764 to the Present*, London, Continuum.

Crow, C.L. 2009. *American Gothic*, Cardiff, University of Wales Press.

Dovey, J. and Kennedy, H. 2006. *Game Cultures: Computer Games as New Media*, Maidenhead, Open University Press.

Freud, S. 1991. *On Metapsychology: The Theory of Psychoanalysis*, Eds J. Strachey and A. Richards, London, Penguin Books.

Giant Sparrow. 2017. *What Remains of Edith Finch*, Annapurna Interactive.

Juul, J. 2010. *A Casual Revolution: Reinventing Video Games and their Players*, Cambridge, MA, MIT Press.

Krzywinska, T. 2002. Hands-on horror. In *ScreenPlay: Cinema/Videogames/Interfaces*, Eds G. King and T. Krzywinska, pp. 206–223, London, Wallflower Press.

McCrea, C. 2009. Gaming's hauntology: dead media in *Dead Rising, Siren* and *Michigan: Report from Hell'*. In *Horror Video Games: Essays on the Fusion of Fear and Play*, Ed. B. Perron, pp. 220–237, London, McFarland & Company.

Mighall, R. 2007. *A Geography of Victorian Gothic Fiction: Mapping History's Nightmares*, Oxford, Oxford University Press.

Miles, R. 2002. *Gothic Writing 1750–1820: A Genealogy*, Manchester, Manchester University Press.

Morris, S. 2002. First-person shooters – a game apparatus. In *ScreenPlay: Cinema/Videogames/Interfaces*, Eds G. King and T. Krzywinska, pp. 81–97, London, Wallflower Press.

Punter, D. 2014. Gothic, theory, dream. In *A Companion to American Gothic*, Ed. C.L. Crow, pp. 16–28, Hoboken, NJ, John Wiley and Sons.

Punter, D. and Byron, G. 2010. *The Gothic*, Oxford, Blackwell.

Savoy, E. 2010. The rise of American gothic. In *The Cambridge Companion to Gothic Fiction*, Ed. J.E. Hogel, pp. 167–188, Cambridge, Cambridge University Press.

Sedgwick, E.K. 1986. *The Coherence of Gothic Conventions*, London, Methuen.

Snyder, S. 2018. The impossible relationship: deconstructing the private space in *Gone Home*, *Journal of Gaming and Virtual Worlds*, 10(1), 7–20. https://doi.org/10.1386/jgvw.10.1.7_1

Sontag, S. 1979. *On Photography*, London, Penguin.

Taylor, L.N. 2009. Gothic bloodlines in survival horror gaming. In *Horror Video Games: Essays on the Fusion of Fear and Play*, Ed. B. Perron, pp. 46–61, London, McFarland & Company.

The Chinese Room. 2012. *Dear Esther*, The Chinese Room and Curve Digital. Available at: https://en.wikipedia.org/wiki/Dear_Esther

The Chinese Room and SCE Santa Monica Studio. 2015. *Everybody's Gone to the Rapture*. Sony Computer Entertainment. Available at: https://en.wikipedia.org/wiki/Everybody%27s_Gone_to_the_Rapture

Part 3

Frolics with Monsters

Chapter 8

Dead Chatty: The Rise of the Articulate Undead in Popular Culture

Bethan Michael-Fox

Abstract

This chapter offers a critical reading of a range of television narratives centred on diverse populations of the articulate dead, including grim reapers (*Dead Like Me*), sort-of-ghosts (*American Horror Story*), zombies (*iZombie*), what appear to be 'just regular dead people' (*The Good Place*, *Les Revenants*) and some other creepy and unusual manifestations of the undead (*Intruders, The Fades*). It suggests that the preponderance of the articulate dead on television is symptomatic of a broader cultural desire to talk both about death and with the dead. It also suggests that there are numerous opportunities to learn from fictional engagement with death and the dead, foregrounding the ways in which televisual narratives can operate to reiterate, critique and engage with social and cultural messages. The chapter takes a playful approach and seeks to distil some key 'self-help' aphorisms that the dead in these series might offer the living about how to approach life, death and everything inbetween, as they tell their audiences to 'look within' to identify the greatest threats to their selfhood, to persevere because 'it's never too late to change', and to 'never forget' the dead and what they might have scarified for the living.

Keywords: Undead; television; self-help; zombies; resurrection; popular culture

Introduction

The presence of the dead in popular culture, as zombies, ghosts, vampires and everything in-between has been a notable and much discussed feature of the contemporary *zeitgeist*. Žižek (2000, 22) has suggested that the last decade of the

Death, Culture & Leisure: Playing Dead, 111–124

doi:10.1108/978-1-83909-037-020201015

twentieth century saw the return of the dead become the 'fundamental fantasy of contemporary mass culture'. Khapaeva (2017) has argued that the 1990s saw the emergence of a 'cult of death' far beyond screen culture, with death making a distinctive mark in fashion, tourism and academia. Davis drives the point home, emphasising the quotidian presence of the dead in contemporary culture and citing two prominent death related cinematic and televisual offerings from the 1990s:

> when the child in *The Sixth Sense* tells Bruce Willis 'I see dead people', it is a stunning cinematic moment, but not a *surprise*. Through seven series, on a weekly basis *Buffy the Vampire Slayer* talked, fought and slept with the undead, and she herself died and returned from the dead. If the haunted child of *The Sixth Sense* had not seen dead people, it would evidently have been because he had not been to the cinema or seen much television recently. (Davis, 2007, 1, emphasis in original)

There appears to be critical consensus that, in popular culture at least, the dead are all around us.

Yet the current preponderance of the dead in popular culture can be understood as signalling the endurance, rather than the emergence, of a concern with the return of the dead. Their proliferation and location may be new but their presence in general is not. As Davis (2007, 1) acknowledges, 'ghosts, the dead and the undead walk among us', but only 'now as much as ever'. The dead have always walked and talked in literature. Arguments positioning the twentieth century as characterised by death denial (Gorer, 1965; Ariès, 1976; Mellor and Shilling, 1993) and those that position a refusal to accept mortality as central to the human psyche, such as Becker's (1973) *The Denial of Death*, have contributed to the arguably still dominant notion across a range of popular discourses and academic disciplines that death is a societal taboo. Yet this is a notion that can be challenged from a range of co-ordinates. Dollimore (2001, 126), alluding to Becker's influential text, has explicitly stated that 'in philosophical and literary terms there has never been a denial of death'. Fuss (2013, 1) has pointed out that the dead could be easily located 'chattering away in poetry' in the twentieth century, undermining 'modernity's premature proclamation of death's demise'. In the twenty-first century death and the dead appear to have become so commonplace that they 'are ever-present and far from being denied, repressed or a societal taboo' (Penfold-Mounce, 2018, 4). Walter (2018) has argued that, in some ways, the dead can now be understood as 'pervasive'. Though some have argued that commonplace exposure to death via twenty-first century screen cultures has further facilitated the denial of 'real' death (Noys, 2005; Bauman, 2004), this chapter will argue that the presence of the dead in popular culture can offer opportunities for meaningful engagement with death and the dead. This is particularly so when the dead are articulate, and when the narratives that house them explicitly provoke practical, philosophical and existential questions about mortality. Louise Winter (2019), who has described herself as one of a 'new breed' of funeral directors in the UK, has written that though 'death is often described as the last taboo', she finds that 'lots of people want to talk about it'. This chapter examines a range of televisual narratives that feature the articulate dead. It suggests that these narratives are emblematic of a desire to talk about death and with the dead, and that in turn they also operate as prompts for doing so.

The televisual narratives under consideration in this chapter have received less critical attention than more dominant zombie and vampire narratives. The articulate dead within them are difficult to classify. They often play with established, though always permeable and contested, boundaries between ghosts, vampires and zombies, signalling existing academic and popular interest in mapping, delineating and exploring differences between the undead that can be seen in projects such as Williams and Schafer's (2013) taxonomy of ghosts and Thompson's (2014) *The Map of Zombies*. Divisions between the undead are arguably particularly fluid on television because of what Jowett and Abbott (2013, xiii) consider to be the 'inherently hybrid nature' of television as a medium. The blending of genres in television has meant that it has been necessary 'to rethink what we mean by horror within a televisual context' (Jowett and Abbott, 2013, xiii) and has allowed for the subversion of tropes associated with representations of the undead. Halberstam (1995, 26) has shown how monsters, too, have always been 'mobile and open to multiple interpretations'. It is their inherent mobility and capacity to imply manifold meanings combined with television's own hybridity that means the debatably monstrous articulate dead have appeared in perhaps unexpected televisual spaces, including suburban comedies (*The Santa Clarita Diet*) and children's shows (*Sesame Street*).

It is perhaps unsurprising and unremarkable that the articulate dead are an important feature of contemporary popular culture. As Fisher (2009, 4) has pointed out, we are in a stage of capitalism that 'subsumes and consumes all of previous history'. Everything has been brought into the market, and not always for the first time. The reanimating of the (already reanimated) talking dead of nineteenth century literature has been a staple of film and television. There are countless adaptations of and homages to Mary Shelley's *Frankenstein* (1818) and Bram Stoker's *Dracula* (1897). The articulate dead are a commodity, and this alone might account for the proliferation of so many examples of them in popular culture narratives. However, many have argued that there is something greater at play in the current dearth of the dead in popular culture, with zombie narratives, in particular, often being positioned as symbolising or negotiating greater socio-cultural anxieties and concerns. The popularity of zombie narratives, in particular in film, has been related to population growth (Luckhurst, 2015), positioned in relation to the terror attacks of 11 September 2001 (Birch-Bayley, 2012; Bishop, 2009) and considered emblematic of humanity's fears about its own capacity for self-destruction (Michael, 2016). When the articulate dead emerge on television they are also open to multiple readings, and a number of potential meanings can be derived from television's more marginal and less easily categorised chattering dead.

As Livingstone (1998, 1) has argued, television 'constitutes a domain in which people ordinarily share experiences of the same complex, "social "messages"'. As such, television is a place where both dominant and marginal ideas about death and the dead can be negotiated. Reed and Penfold-Mounce (2015, 136) have suggested that engagement with death and the dead through popular culture can be particularly meaningful, prompting reflection on and engagement with 'sweeping sociological themes'. They argue that the television series *The Walking Dead* can be understood as social-science fiction, inspiring 'the sociological imagination amongst its largely non-scholarly audience' (Reed and Penfold-Mounce, 2015, 136). Where *The Walking Dead* franchise is predominantly focussed on the living, the narratives discussed in this chapter shift the focus from the living to the dead. According to Baldick (1990, 45), it was giving the creature in *Frankenstein* 'an articulate voice' that was 'Mary Shelley's most important subversion of the category of monstrosity'. The televisual narratives discussed here further this legacy by giving the dead, as Shelley did, a voice, an opportunity to speak and a central role in the narrative. In doing so, they offer a space in which to engage with ideas about death, consider what day-to-day life might be like for the dead and to imagine what the dead might have to say.

In the spirit of this collection, this chapter takes a playful approach to structuring the analysis of the texts examined here. The discussion of a range of examples of televisual texts that feature an articulate dead is themed in a way that emphasises what we might learn from these chatty dead, or what 'social "messages"' (Livingstone, 1998, 1) these narratives offer and negotiate. As evidence mounts that a shift away from the reiteration of the death denial thesis is occurring, these series are open to being read as offering self-help messages that centre on what we can learn from engagement with death and the dead. Self-help, self-care and self-improvement are all themselves popular elements of the contemporary *zeitgeist*, and the world of self-help and the undead collide in the *Zen of Zombies* book series, which includes *The Vampire Seduction Handbook: A Guide to the Ultimate Romantic Adventure* (Ballion and Bowen, 2009) and *Z.E.O: A Zombie's Guide to Getting A(Head) in Business* (Kenemore, 2009). See also, for example, Buckely's discussion of Zombie fitness apps in this volume. This chapter is structured into three aphorisms that encapsulate some of the central messages that the dead in these narratives can be understood to deliver. These are:

1. Look within (*American Horror Story: Murder House* and *Intruders*);
2. It's never too late to change (*iZombie, Dead Like Me* and *The Good Place)*; and
3. Never forget (*Les Revenants* and *The Fades*).

Giroux (2011, 102) has argued that in recent decades 'making the world a better place has given way to collective narratives about how to survive alone in a world whose destruction is just a matter of time'. As such, he suggests, 'death, fear, and insecurity' have come to take precedence over 'crucial questions about what it means to apprehend the conditions to live a good life in common with others' (Giroux, 2011, 102). The narratives under discussion here can be read as working against this grain, offering messages from the dead that encourage self-awareness, perseverance and remembrance.

Look Within

The imperative to look within as a means for self-improvement, to enhance self-awareness and to find happiness has roots in a range of religious, spiritualist, moralistic, psychoanalytic and, more recently, neoliberal discourses. In the Gothic tradition, where the dead have often been found to speak, 'terror and horror have depended on things not being what they seem' (Botting, 1996, 170). This includes the self and is perhaps one reason why the Gothic has been so susceptible to psychanalytic readings that emphasise the power of the unconscious. Halberstam (1995, 27) argues that as the Gothic became postmodern, attempts to 'identify the monster and fix the terms of his/her deformity' gave way to warnings to 'be suspicious of monster hunters, monster makers, and above all, discourses invested in purity and innocence', as a concern with the monstrosity of the other was overtaken by an interest in the 'need to recognize and celebrate our own monstrosities'. The series discussed in this section can be easily described as examples of Gothic television, advancing a tradition of emphasising internal threats and the monster within. The first season of the highly successful television series *American Horror Story* and the only season of the far less popular series *Intruders* are also examples of recent narratives focussed on the return of an articulate dead suggesting that, in a cultural climate of high anxiety and frequent reminders of the requirement to be vigilant to external threats, we might also be prudent to focus our attention on the threat within.

The first season of *American Horror Story*, given the subtitle *Murder House* by fans, is populated by a diverse group of self-aware dead people. In the cartography of the dead, they are closest to ghosts, though not easily categorised as such as they are corporeal and capable of most of what they were when alive, including touch, sex, murder and procreation with the living. Unlike most zombies, they retain their personalities. Unlike most vampires, the dead in this particular series of *American Horror Story* are only ever emotionally dependent on the living. The dead are confined by an inability to leave the supernaturally powerful house in which they die. In this sense, they are ghostly. They are prevented from experiencing or partaking in life in the outside world, from ageing, sleeping or from fully shuffling off their mortal coil. Anyone who dies inside the house remains there forever. The season is focussed on the Harmons, whose lives have been marred by loss (in the form of the wife Vivien's late term miscarriage) and betrayal (in the form of the psychoanalyst husband Dr. Harmon's affair with a young student). They share their new Los Angeles home, where they hope to overcome their past traumas and find happiness, with a host of dead occupants dating back to 1920s, including a victim of the real Black Dahlia murders in 1940s Los Angeles (*American Horror Story* often blurs the boundaries between the historical and the imagined). Many of the dead mean the living no harm. Those who do are often shown to be violent as a consequence of nurture rather than nature, possessing complex psychological motivations they can articulate in therapy with Dr Harmon. Psychology and psychoanalysis are central themes in the narrative, and the season has already been considered in relation to psychoanalytic concepts. Keetley (2013, 90) argues that it 'incarnates the death drive, and, as such, it is one of the most unrelentingly pessimistic cultural texts of the contemporary moment'. Yet compared to other seasons of *American Horror Story,* this one ends on what might be deemed a relatively hopeful note, at least for the Harmon family, who by the end of the season are all dead and reunited as a happy family with their new (dead) baby.

As Keetley (2013, 91) also points out, the series manifests 'almost every conceivable Gothic convention'. Characters display a compulsion to repeat, there are various incidences of uncanny returns and doubles and psychological rather than physical torments are emphasised in a domestic setting. Most of the dead characters are haunted by their past, having suffered or inflicted great pain. Often, their fears are revealed not to be the dead that surround them but the emergence or re-emergence of that which is already within them. One example of the season's apparent message to 'look within' can be seen in the character Violet. Violet is the unhappy teenager of the Harmon family who has discovered that her boyfriend Tate is dead. She fears he is trying to kill her. He leads her into the depths of the house, where she is shocked to discover that he has been hiding the utmost abject – her own rotting corpse (Kristeva, 1982). Having intentionally overdosed two weeks earlier and woken as one of the house's dead occupants, Violet had allowed herself (and the audience) to believe that she had been saved by Tate and survived her own suicide attempt. Violet perceived the threat to her life to be external and embodied in the dead Tate. Focussed on an outside threat, she struggles to acknowledge that her own internal conflict and suffering has already brought an end to her life.

The series *Intruders* depicts a less explicit return of the articulate dead, though there is a similar emphasis on the threat within and on suicide. In this series, dead members of the elite secret society *Qui Reverti* (they who return) are able to achieve immortality by inhabiting the bodies of the living. Within each person there can be two potential selves, the currently living one and the dormant dead one (or vice versa). *Qui Reverti* arrange for their own awakening inside of new bodies, the current occupants of which experience drastic personality changes as they are deposed by the dead within them. In this sense, the plot is not dissimilar to that of Jordan Peele's *Get Out*. *Intruders* depicts a world in which there can be dead others with us, literally inside of us, throughout our lives – a nod, perhaps, to the psychoanalytic concept of the unconscious 'discovered' by Freud (1915) and the idea that, as Kristeva has written, we are *Strangers to Ourselves* (1991). One character, who is made able to see both the dormant and awakened people living inside of bodies, is driven to suicide when he sees that his infant child is not his infant child but someone else entirely, an old soul who has forced their way into his child's body. Throughout the series, the return of the dead remains a well-hidden conspiracy. There is no global apocalypse to contend with in either *Intruders* or in *Murder House*. Both series emphasise internal rather than external threats, play with boundaries between human and monster and urge their audiences to 'look within'. Certainly, a similar imperative can be identified in many other cultural texts premised on the return of the dead, articulate or otherwise, including some of the most popular post-apocalyptic narratives of the early twenty-first century. *The Walking Dead*, for example, plays with this idea given the zombie virus in the franchise is 'already inside' of all humans, waiting to emerge after death regardless of how they die. What is more unique in *Murder House* is the message seemingly conveyed by the finally happy Harmons at the end of the season: being dead might actually solve your problems. In *Intruders*, where the opportunity to survive death is readily available to society's most elite, the central message might not be one about focussing on the individual self, but rather one prompting a deeper consideration of the broad ranging consequences of engrained structures of privilege as they prevail even in death.

It's Never too Late to Change

The hope that you might yet change for what you perceive to be the better – physically, psychologically, emotionally, at work or in your home life – is the mainstay of self-help. In contemporary narratives of the articulate dead, even zombies are capable of change. Reed and Penfold-Mounce (2015, 129) have pointed out that 'rehumanized zombies' have become a 'growing theme' in recent years. They can be located in literature, film and television. *Shaun of the Dead* features at its conclusion footage of rehabilitated zombies and in the novel turned film *Warm Bodies,* love makes a zombie human again. The television series *In The Flesh* focusses on rehumanised zombies and explores themes of redemption, discrimination and self-acceptance as the dead come to terms with their new existence in the face of pressure to conform when the living encourage them to 'blend in' with make-up and contact lenses. In the narratives discussed in this section, the message that it is never too late to change tends to centre on the notion that just because you have died, it does not mean you cannot become a better person. The three examples examined here are *iZombie*, *Dead Like Me* and *The Good Place*. It is notable that all of these series have a young, slim, white and blonde woman as the central character. Though they are perhaps symptomatic of the pervasive lack of diversity in some areas of popular culture, they are perhaps also indicative of the tendency for so much transformational self-help to be aimed both at – and more problematically at becoming more *alike* – this particular demographic. Yet each series can also be read as critiquing and playing with the expectations placed upon young women, as none of the central characters succeeds in living up to, in life or in death, the ideals they seem to be measured against.

iZombie, a television adaptation of a comic book series, follows the undead life of a committed medical resident turned zombie who works with the police to solve murders. Crime solving dead people are well established in television, featured in the comedy *Pushing Daisies*, which centres on a character able to revive corpses momentarily in order to ask questions of the dead, and the 2000 remake and original of Dennis Spooner's 1969 British comedy-drama *Randall and Hopkirk (Deceased)*, in which a dead detective returns as a ghost to solve his own murder as well as other crimes. *iZombie* adds to this tradition but also plays into a range of ideas about transforming the self and 'having it all'. The central character Liv is 'all work and no play'. She finally makes the time to attend a party and promptly gets scratched by a zombie. Zombie-ism in this series is the product of mixing a tainted batch of a synthetic drug named Utopium with an energy drink, making zombies well suited to the inexhaustible speed of 24-hour twenty-first century living. In order to disguise her deceased status Liv gains employment in a coroner's office where she is able to consume the brains of unidentified corpses (relatively) ethically, disguising her post-mortem pallor as a 'Goth' makeover. When consuming the brains of the dead she also consumes their memories, adopts their personality traits and is compelled to solve their murders. Liv gives a whole new meaning to Polhemus's (1998) notion of the postmodern 'supermarket of style' as she tries out the different personalities of different dead people and learns about herself along the way. Liv can be a different person every day, experimenting with being a party-girl, a dominatrix, a geek, a kleptomaniac, a pretentious artist and a sociopath amongst others, with relatively few consequences to speak of. Liv does not try out these many different selves by choice. Rather, she must experience them in order to retain her own self-hood and personality. If she fails to consume brains, she will deteriorate into the archetypical zombie known in the series as 'Romeros' in a nod to George A. Romero. Her adventures into different personalities imply both the ever-permeable nature of the monster and of the self. Liv's ultimate challenge is perhaps that seemingly thrust upon so many young women, as she struggles to stay 'true to herself', whatever that might mean, when faced with so many competing possibilities for who she might be.

Dead Like Me is focussed on a listless young woman who, having been exceptionally unmotivated in life, is given new opportunities in death to make friends, make amends and even have sex for the first time. The dead in this series are also put to use, this time as grim reapers. They appear physically different to their loved ones so cannot continue life in any 'normal' way. Rather, they are able to witness the suffering and loss experienced as a consequence of their own demise. The series has a reflective tone and offers an exploration of grief as well as prompting audiences to question what they might do differently with a second chance at life. The central character, 18-year-old college dropout George is crushed to death by a toilet seat that falls to earth from a disorbiting space craft. She carries her generally apathetic attitude into death and her views are expressed in a voice-over commentary where she makes observations like: 'as childhood traumas go nothing beats the realisation that everything dies, including you' (s.1, e.1) and 'life sucks and then you die...and then it still sucks' (s.2, e.1). Throughout *Dead Like Me* George learns lessons she did not learn when she was alive. She is mentored by a dead father figure, Rube, whose death took him away from his own daughter. All of the deceased in the series remain attached to and preoccupied by the same concerns that consumed their lives and though in death they have even less control or capacity to affect change, they are able to reflect on the mistakes that, in hindsight, they feel they made. Dillman (2014, 85) has described such characters as George as women who are 'dead but not gone', who begin to exercise agency only after their death. Yet she also points out that George's dead agency is limited to nostalgia and retrospection. The series expresses a particular concern with tensions between structure and agency, emphasising that an individual's capacity for change will only ever be minimal, limited by the chance nature of life and death, by others and by bureaucratic systems that dictate, in this instance, what happens to you after death – it was only by a combination of immovable bureaucracy and random chance that George ended up a grim reaper. In this series, it is never too late to change. However, though presented as ultimately worthwhile, any change is small, and any differences made are minimal, private and domestic. With its quiet focus on grief, *Dead Like Me* survived for only two seasons and a direct-to-DVD film.

The exceedingly more popular *The Good Place* imagines, rather than the world with dead people in it, an afterlife full of dead people. In this series, which explicitly plays with Sartre's most-ill-quoted aphorism 'hell is other people', the dead are in heaven/hell. What critic Andrew Street (2018) calls 'ratings gold' seem to have been 'found in a sitcom about the afterlife, with regular references to Immanuel Kant, David Hume and Aristotle'. The talkative dead in this series, in particular the central character Eleanor, have been terrible people in life. Eleanor is constantly compared to the 'other' Eleanor whose place it seems she has taken in heaven. The 'other' Eleanor was philanthropic, generous and kind. The real Eleanor is selfish, rude, dishonest, promiscuous and swears constantly. In this sense, *The Good Place* perhaps plays with ideas about what a 'good' woman is and is not. Yet self-improvement is still on the agenda, as Eleanor must be guided towards becoming a 'good' person. Like in *iZombie* and *Dead Like Me*, the dead are given the opportunity to learn about themselves and potentially change the course of their own futures. One of the playful twists in this series is that the character spouting the philosophy lessons and teaching Eleanor her new ways, Chidi, turns out to be just as awful as everyone else – he talks the talk but has failed to walk the walk of ethical living. In some ways, these three series offer representations of agentic, unconventional and independent young women as central protagonists. As mentioned earlier, young women are often the traditional 'targets' of self-help and in these series, they do seem to be 'helping themselves', though often with explicit guidance from a man. Yet in the strikingly similar appearance of the central characters and in the focus of all three series on the 'transformation' of young women, each raises important questions about why, when they all imply that change is possible, mainstream popular culture continues to demonstrate a preoccupation with both narrow constructions of womanhood and with a woman's responsibility to improve herself, even when she is dead.

Never Forget

The two aphorisms offered so far by the articulate dead of contemporary television are focussed on the self – on looking within, and on being committed to changing yourself. Yet a range of recent television series premised on the return of a relatively mundane dead seem to emphasise the importance of remembering as they explore the responsibility of the living towards the dead, shifting the focus from how the living might improve their lives to how the living might reconsider their treatment of the dead. The series discussed in this section are perhaps most easily read in terms of the critical practice of hauntology, the roots of which lie in Derrida's *Specters of Marx* (1994). Derrida (1994, xviii) writes, in a kind of self-help dictum of his own, that we must learn to 'live *with* ghosts' as a means to live, if not 'better', then 'more justly'. The series considered in this section imagine what it might look like to live with the dead were they to return and share a similar emphasis as Derrida in their concern with responsibility and the need for inclusive, just, and critical remembrance.

Les Revenants is based on a 2004 French film of the same name. It has been released with subtitles by the *BBC* and adapted into the US series *The Returned,* which was released a year after the remarkably similar US series *Resurrection,* based on a novel also titled *The Returned* (2013) by Jason Mott. The Australian series *Glitch* also has an extremely similar premise and is perhaps indicative of a broad fascination with the notion of a sudden, un-apocalyptic return of the ordinary dead. Blanco and Peeren (2010, xiv) have suggested that one reason for the current preponderance of ghosts and the ghostly in popular culture might be that both offer apt metaphors for marginalised populations who 'haunt the everyday, living on the edge of visibility and inspiring a curious mix of fear and indifference'. Series that bring back the dead, often those marginalised in life, and posit them as marginalised populations can be seen to utilise them to negotiate similar concerns. *Les Revenants* is less obviously a social commentary on historical and contemporary marginalisation than *Resurrection* or *Glitch,* which deal more explicitly with American slavery and Australian colonialism, respectively. Yet *Les Revenants* is also markedly different from more traditional zombie apocalypse narratives in that it has a 'brooding and meditative' tone which makes it clear it is 'less about gore than grief' (Mangan, 2014). The series is full of gothic tropes – uncanny returns, doubles and dark, mountainous landscapes. Luckhurst (2015, 188) has commented that 'the breakout success' of *Les Revenants* serves to suggest that this kind of

> subtly disturbing narrative of the dead returning to their families and lovers as brute physical presences, markers of stalled mourning or melancholic denial, can twist away from the splatter-gore aesthetic that has predominated since Romero.

As Luckhurst (2015) implies, a Freudian reading can be easily applied here. Those who return seem to be those for whom someone living has been unable to mourn 'successfully' and who remains in a state of melancholia (Freud, 1915). The past coming back, the unreliability of memory and the tendency to imagine people as you wanted them to be, rather than as they were, are all central to *Les Revenants.* The series can also be seen to challenge the dominance of self-help messages for the bereaved that emphasise the importance of 'moving on'. It seems to support instead the message delivered by Nora McInerny (2018) in her popular *TedTalk* about bereavement, which emphasises that it is possible to, rather than 'move on' from the dead, 'move forward' with them.

The British series *The Fades* focusses more on cultural than individual memory, engaging with ideas about the consequences of war and the living's responsibility towards the dead. In this series, the 'natural' process of dying has been disrupted by technological advances, modern warfare and the invention of concrete. Rather than 'ascending' to some other plane, some of the dead remain on earth, gradually decaying and unable to touch, feel or communicate with the living. This 'hell on earth' is indiscriminate, with those who remain chosen at random rather than because of anything they have done in life to deserve such a fate. As the character Neil explains (s.1, e.1), the unfairness of life is reflected in death: 'Life has famine, illness, shittiness...death is similarly crap'. The decaying dead find a way return to a corporeal state by consuming the living. The disruption of the 'natural' process of dying is positioned as having begun in 1943 during World War II, and in this sense the series seems to play in some ways into the notion of the twentieth century as one of death denial. The mass casualties of the war are explained as having broken the ladder of ascension to another plane and the undead are positioned as victims of human atrocity. Whereas in *Les Revenants* the dead seem to return to those who focussed most on the act of remembering, the dead in *The Fades* have existed for decades without regard, the living ignorant of their circumstance. When they do come back to corporeal form, they bring history lessons with them and remind the living of the consequences of the past. Unlike many common self-help aphorisms that encourage individuals to focus on the future and let go of the past, the message in these series seems to be that the power of memory should not be underestimated. It is the responsibility of the living to remember, but also to remember critically, and to remember that remembering itself is an inherently unstable, unreliable and political act.

Conclusion

To some extent, the proliferation of the dead in twenty-first century popular culture is merely a consequence of the general proliferation of everything. Yet much recent engagement with the dead in popular culture might also be understood as timely. Will Self (2013) has pointed out that it is perhaps surprising that 'in our modern secular and avowedly inclusive society we have wilfully allowed the dead to be so gagged', or that 'in an era when every minority is, at least in theory, listened to, we have turned our backs on the great majority and rendered them silent'. Arguably, it is about time the dead were given more opportunity to talk. The presence of the dead in popular culture narratives may indicate both a broader desire to hear from the dead, and to engage with philosophical, practical and existential questions about mortality. Often the concerns of the dead in the series discussed here are also the concerns of the living, and the dead are utilised to explore a range of social and culture issues that are not explicitly death related, as the dead and their representation become vehicles for a wide range of potential meanings. Yet the popularity of television series about the dead, from explicitly philosophical comedies to irreverent ones and from melancholic and haunting dramas to gore filled horror, suggest some level of interest in imagining the lives of an articulate dead and on shifting the focus from the living to the dead. These series might be understood to offer opportunities for audiences to begin conversations about death and the dead, and they might also be understood as a challenge to the much cited and repeated tenets of what has come to be known as the death denial thesis. Like all cultural texts, they are open to multiple interpretations. Many of the popular culture narratives explored in this chapter play with, subvert and blur the permeable and constructed boundaries between categories of the undead and between humans and monsters. They can also be read as exploring loss, grief and memory and as susceptible to psychoanalytic, feminist and other readings. As the articulate dead continue to make their voice in contemporary culture heard, they raise questions, suggest answers and prompt conversation – perhaps even offering a little self-help.

Interactions and Sensoria

Davis, C. 2007. *Haunted Subjects: Deconstruction, Psychoanalysis and the Return of the Dead*, Basingstoke, Palgrave MacMillan.
Fresco, V. (Creator). 2016–2019. *The Santa Clarita Diet*, Netflix.
McInerny, N. 2018. We don't 'move on' from grief, we move forward with it, *TEDWomen*, November. Available at: http://bit.ly/2pm8QPf
Mitchell, D. (Creator). 2013–2014. *In the Flesh*, BBC Three.
Penfold-Mounce, R. 2018. *Death, the Dead and Popular Culture*, Bingley, Emerald.

References

Ariès, P. 1976. *Western Attitudes toward Death: From the Middle Ages to the Present*, Translated by P.M. Ranum, London, Marion Boyars.
Ayres, T., Fox, L. and Hill, A. (Creators). 2015. *Glitch*, Sydney, Matchbox Pictures.
Baldick, C. 1990. *In Frankenstein's Shadow: Myth, Monstrosity, and Nineteenth-century Writing*, Oxford, Clarendon Press.
Ballion, L.R. and Bowen, S. 2009. *The Vampire Seduction Handbook: A Guide to the Ultimate Romantic Adventure*, New York, NY, Skyhorse.
Bauman, Z. 2004. *Wasted Lives: Modernity and its Outcasts*, Cambridge, Polity.
Birch-Bayley, N. 2012. Terror in horror genres: the global media and the millennial zombie. *Journal of Popular Culture*, 45(6), 1137–1151. https://doi.org/10.1111/jpcu.12001
Bishop, K. 2009. Dead man still walking: explaining the zombie renaissance, *Journal of Popular Film and Television*, 16(1), 16–26. https://doi.org/10.3200/JPFT.37.1.16-25
Blanco, M.d.P. and Peeren, E. 2010. Introduction. *Popular Ghosts: The Haunted Spaces of Everyday Culture*, Eds M.d.P. Blanco and E. Peeren, pp. xi–xxiv, London, Continuum.
Botting, F. 1996. *Gothic*, London, Routledge.
Bryan, F. (Creator). 2003–2004. *Dead Like Me*, ABC Studios.
Campillo, R. (Director). 2004. *Les Revenants*, Haute et Court and Gimages Développement (105 minutes)[1].
Cooney, J.G. and Morrisett, L. (Creators). 1969. *Sesame Street*, Sesame Workshop and the Children's Television Workshop.
Cuse, C. (Developer). 2015. *The Returned*, A&E Studios.
Davis, C. 2007. *Haunted Subjects: Deconstruction, Psychoanalysis and the Return of the Dead*, Basingstoke, Palgrave Macmillan.
Dillman, J.C. 2014. *Women and Death in Film, Television, and News: Dead but not Gone*, Basingstoke, Palgrave Macmillan.

[1]The film was given the international title *They Came Back* and also released as *The Returned* in the UK.

Dollimore, J. 2001. *Death, Desire and Loss in Western Culture*, New York, NY, Routledge.
Falchuk, B. and Murphy, R. (Creators). 2011. *American Horror Story: Murder House*, FX.
Fisher, M. 2009. *Capitalist Realism: Is There No Alternative?*, Winchester, Zero Books.
Freud, S. 1915. The unconscious. In *The Standard Edition of the Complete Psychological Works of Sigmund Freud Volume XIV (1914–1916)*, Ed J. Strachey, pp. 159–209, London, Vintage.
Fuller, B. (Creator). 2007-2009. Pushing Daisies. ABC Studios.
Fuss, D. 2013. *Dying Modern: A Meditation on Elegy*, Durham, NC, Duke University Press.
Giroux, H.A. 2011. *Zombie Politics and Culture in the Age of Casino Capitalism*, New York, NY, Peter Lang.
Gobert, F. (Creator). 2012–2015. *Les Revenants*, Haut et Court.
Gorer, G. 1965. The pornography of death. In *Death, Grief and Mourning in Contemporary Britain*, Ed. G. Gorer, pp. 169–175, London, The Cresset Press.
Halberstam, J. 1995. *Skin Shows: Gothic Horror and the Technology of Monsters*, Durham, NC, Duke University Press.[2]
Higson, C. (Writer and Producer). 2000–2001. *Randall and Hopkirk (Deceased)*, Working Title Television.
Jowett, L. and Abbott, S. 2013. *TV Horror: Investigating the Dark Side of the Small Screen*, London, I.B. Taurus.
Keetley, D. 2013. Stillborn: The entropic gothic of American horror story, *Gothic Studies*, 15(2), 89–107. https://doi.org/10.7227/GS.15.2.6
Kenemore, S. 2009. *Z.E.O: A Zombie's Guide to Getting A(head) in Business*, New York, NY, Skyhorse.
Khapaeva, D. 2007. *The Celebration of Death in Contemporary Culture*, Ann Arbor, MI, University of Michigan Press.
Kristeva, J. 1991. *Strangers to Ourselves*, New York, NY, Colombia University Press.
Kristeva, J. 1982. In *Powers of Horror: An Essay on Abjection*, Translated by L.S. Roudiez, New York, NY, Columbia University Press.
Levine, J. (Director). 2013. *Warm Bodies*, Mandeville Films (98 minutes).
Livingstone, S. 1998. *Making Sense of Television: The Psychology of Audience Interpretation*, 2nd ed., London, Routledge.
Luckhurst, R. 2015. *Zombies: A Cultural History*. London: Reaktion.
Mangan, L. 2014. The returned – Box set review, *The Guardian*, February 20. Available at: http://bit.ly/2onRf9c
McInerny, N. 2018. We don't 'move on' from grief, we move forward with it, *TEDWomen*, November. Available at: http://bit.ly/2pm8QPf
Mellor, P.A. and Shilling, C. 1993. Modernity, self-identity and the sequestration of death, *Sociology*, 27(3), 411–431. https://doi.org/10.1177%2F0038038593027003005
Michael, B. 2016. The return of the dead: Fears and anxieties surrounding the return of the dead in late postmodern culture. In *There's More to Feat than Fear Itself: Fears and Anxieties in the 21st Century*, Eds I. Dixon, S.E.M. Doran and B. Michael, pp. 31–44, Oxford, Inter-Disciplinary Press.
Mitchell, D. (Writer and creator). 2013–2014. *In the Flesh*, BBC Three.
Morgan, G. (Developer). 2014. *Intruders*, BBC America.
Mott, J. 2013. *The Returned*, Toronto, Mira Books.
Noys, B. 2005. *The Culture of Death*, Oxford, Berg.
Peele, J. (Director). 2017. *Get Out*, Blumhouse Productions (104 minutes).
Penfold-Mounce, R. (2018). *Death, the Dead and Popular Culture*, Bingley, Emerald.

[2]Although some still refer to Halberstam as Judith, they prefer to go by the name Jack. See: http://bit.ly/2pjCqoD for further information.

Polhemus, T. 1998. In the supermarket of style. In *The Clubcultures Reader*, Eds S. Redhead, D. Wynne and J. O'Connor, pp. 130–133, Oxford, Blackwell Publishers.
Reed, D. and Penfold-Mounce, R. 2015. Zombies and the sociological imagination: the walking dead as social-science fiction. In *The Zombie Renaissance in Popular Culture*, Eds L. Hubner, M. Leaning and P. Manning, pp. 124–138, Basingstoke, Palgrave MacMillan.
Schur, M. (Creator). 2016–2019. *The Good Place*, Fremulon.
Self, W. 2013. It's always the others who die, *A Point of View, BBC Radio 4*, 8 December. Available at: https://bbc.in/2BRoTXX
Shelley, M. 1992 [1818]. *Frankenstein: Or, the Modern Prometheus*, London, Penguin.
Stoker, B. 2004 [1897]. *Dracula*, London, Penguin.
Street, A.P. 2018. The good place: How a sitcom made philosophy seem cool. *The Guardian*, 29 January. Available at: http://bit.ly/345kzR9
Thomas, R. and Ruggiero-Wright, D. (Developers). 2015–2019. *iZombie*. The CW.
Thorne, J. (Creator and writer). 2012. *The Fades*, BBC Three.
Thompson, J. 2014. The map of zombies, *Kickstarter*. Available at: http://bit.ly/32Ukj7p
Walter, T. 2018. The pervasive dead, *Mortality*, *24*(4), 389–404. https://doi.org/10.1080/13576275.2017.1415317
Williams, R. and Schäfer, H. Eds 2013. *Dis Manibus: A Taxonomy of Ghosts from Popular Forms*, York, Information as Material.
Winter, L. 2019. Funerals should be personal, but new coffin colours aren't the answer. *The Guardian*, 22 January. Available at: http://bit.ly/2onAsD9
Wright, E. (Director). 2004. *Shaun of the Dead*, WT^2 (100 minutes).
Zelman, A. (Developer). 2014–2015. *Resurrection*, ABC Studios.
Žižek, S. 2000. *Looking Awry: An Introduction to Lacan through Popular Culture*, Amherst, MA, Massachusetts University Press. https://www.amazon.co.uk/Looking-Awry-Introduction-Jacques-through/dp/026274015X

Chapter 9

The Slender Man: The Internet's Playful Creation of a Monster

Vivian Asimos

Abstract

The Slender Man, an online monster born to the internet in 2009, loves to live in the boundaries, specifically the boundary between the digital and the non-digital worlds. This chapter seeks to explore the full engagement of the community with the myth, analysing it structurally to understand the way, in which the narrative's construction reflects the relationship of monster to society and society to itself. The sincerity in which the narratives are told, at first appearing intense, is actually a form of play, revealing a boundary blurring between play and non-play as well. While fully engaging in play, the community's structured narrative is more than this: pulling them into a trapped world in which they play with death in the digital. The triadic structure of the Slender Man narratives reveals an anxiety to the community and their place within broader non-digital worlds.

Keywords: Slender man; structure; digital mythology; digital community; monster; creepypasta

Introduction

The Slender Man is found in a variety of cultures around the world. He is known as *Fear Dubh*, the Black Man, in Scotland, where he lingers in the forest and haunts lonely paths through the woodlands. In Germany, he is known as *Der Großmann*, or the Tall Man, where he prowls the Black Forest. Wherever he is, he leaves violence and terror in his wake. An all-too thin and all-too tall fairy who appears, at first glance, like a man in a suit, that makes children disappear in the middle of the night. His mythology spread to the United States where he took

Death, Culture & Leisure: Playing Dead, 125–137

doi:10.1108/978-1-83909-037-020201016

blame for missing children and mysterious fires which destroyed records of his presence.

Or, at least, that is how the stories go.

In 2009, the *Something Awful* online forums gave birth to the Slender Man. Images were altered to insert the Slender Man's presence, making him linger menacingly behind children.[1] Narratives were produced which mimicked newspaper articles, and government documents, to create avenues of proof for the Slender Man's existence. Through the process of organic communal narrative creation, the internet created a mythology.

The online community engaged with the myth directly, in the way the narratives and myths are told and discussed. The narratives themselves hold such a sincerity, which may appear either unnecessary or misplaced. But this sincerity is an act of play – wherein users, as both storyteller and audience, play with both the story and their own sincerity as an engagement with a mythology of death. The stories themselves also play – they play with the boundaries between digital and non-digital worlds. The structure of the mythos demonstrates the relationship of the monster to the digital society, as well as the society's relationship to itself. The audience and storyteller, sometimes as one and the same, become trapped in their own play with the monster's death in the digital.

My research on the Slender Man formed the basis of a lengthy study of narratives on the Slender Man specifically, as well as other virtual horror narratives and their derivatives; a portion of the findings will be summarised here. The research itself was not limited to the *Something Awful* forum posts, though these formed much of the base study which I found repeated in the other narrative formats. What will be summarised here is a fuller analysis, though the examples will be selected from the *Something Awful* forums in order to keep the conversation as succinct as possible. I will sum up what was elicited and detail a bit more specifically the connection and view of the digital self to the digital community which we can begin to understand.

The Slender Man Mythology

The Slender Man, at first glance, appears to be a tall man wearing a suit. At second, things start to change. His height is inhuman, his limbs strangely proportioned, and his suit appears almost like a second skin. His face, at times, is monstrous, and at other times, entirely without facial features. Sometimes, he has strange appendages which sprout from his back – sometimes tentacles, sometimes tree limbs, and at other times, just shadowed masses.

What makes the Slender Man special, and especially monstrous, is that there is nothing certain about it. His appearance changes from story to story as much as his purpose. As each creator brought the narrative into their own creative thought, the narrative shifted to accommodate its new presence. The community

---

[1]A more detailed overview of these sources and the images can be found in Asimos (2019a).

in its early life decided that set canon should fade in favour of individual creativity. The result was a Slender Man who became known for being incomprehensible – a monster whose victims could not predict his movements at all. As an initial architect of the mythos described, the Slender Man became 'beyond our comprehension' (TheRiffie, 2009).

The Slender Man's story is told through a variety of mediums. The first were images where users digitally manipulated photos to add the Slender Man into it. There were also written narratives, though these were often written to emulate what I term 'proof' narratives, for example, newspaper articles and government emails. The story grew from the *Something Awful* forums to other virtual narratives, including web videos (Marble Hornets, 2009–2014) and video games (Parsec Productions, 2012; Blue Isle Studios and Parsec Productions, 2015;).

The users play with folkloric inspiration and the concept of mythology itself. They play with the shadowy boundaries between reality and fiction, letting him linger in the mid-way point. None of the users who create and shared narratives on *Something Awful* 'believed' in the absolute reality of the existence of the Slender Man. Rather, they play with the elements of belief and reality, to push these normally concrete ideas while remaining in the grasp of fiction. What is at the core of this narrative, is play.

Throughout much of its use in anthropology, play has been inherently associated with childish behaviour – children are the ones who play. The adult forms of play are not 'play', but 'sport' or 'ritual' (Hamayon, 2016). In religious studies, the connection of religion to play has struggled to find ground (Raj and Dempsey, 2010), due to a similar understanding of play being associated with frivolity. Often, at the heart of this issue, is the oft conceived strict dichotomy between play and work, which means that the play becomes less important (Malaby, 2009).

When anthropologists do attempt to demonstrate how play can be meaningful, the term 'serious play' is utilised (Luhrmann, 1989; Ortner, 1999). 'Serious play', like Geertz's (1973) 'deep play', is used when the scholar feels the need to make play more relevant, though the need to distinguish the particular play as different from other forms of frivolous play, still holds up the dichotomy of play as both insignificant and unimportant (Malaby, 2009).

This digression on play and mythology is important to keep in mind in relation to the Slender Man. Users who build his narrative are fully aware of the Slender Man's fictional status – if I had asked any of my participants if they believed the Slender Man was real, I would have been laughed away: they do not believe in his physical presence. However, if we asked them if the Slender Man means something significant to them, or if his narratives have influenced their lives, the answer would be very different. It is this relationship we seek to uncover here.

Analysing the Slender Man

Mythology can, and has been, defined in a multitude of ways (see Malinowski, 1948; Frazer, 1994; Lincoln, 1999; Tylor, 2010; and for introduction on the subject see Miles-Watson and Asimos, 2019). Adopting Seth Kunin's definition, I consider myth as that which a community, or an individual, uses to structure their understanding of themselves and the world around them (Kunin, 2003; Asimos, 2019b). This definition allows for the community to determine meaning, rather than an outside academic making the distinctions for them. This definition also relieves the issue of a dichotomy between truth and fiction that many more traditional definitions of myth have. William Bascom's distinctions between folklore, legend and myth contrast folklore as that which was regarded as 'fiction' with the mythic conceptions of what are 'considered to be truthful accounts' (Bascom, 1984 9). This also allows for contemporary narratives to be seen in the important lights many fans put them in.

My primary case studies from the *Something Awful* forum attempted to gather a variety of types of narratives together. As mentioned previously, the narratives on the *Something Awful* forum mimicked many different forms of what is typically considered as 'proof', that is, photographs, newspaper articles, emails and even government documents. By gathering narratives from different times in the forum thread's life, as well as different types of narrative telling. My primary choices were manipulated images or photographs and their accompanying captions by the first user to have submitted a Slender Man narrative, Victor Surge. Due to the much shorter nature of these, and how they often flowed into one another, I chose four different posts by Victor Surge (2009b, 2009c, 2009d), which is comprised of six photographs . A little later in the forum's lifespan, a user by the name of Mr. 47 posted a written narrative which took the shape of a newspaper article (Mr. 47, 2009). User Aleph Null, too, wrote a narrative which mimicked that of an esoteric book (Aleph Null, 2009). The last case study choice combined written text and photographs, by mimicking both a police and psychiatric report on children who witnessed a kidnapping, paired with pictures of the Slender Man drawn by the children (genesplicer, 2009).

Of these narratives, three primary mythemes, or the smallest element of a myth, surfaced, which have been grouped with similar or related mythemes in columns:

There is much to be said in Table 1, including Slender Man's lack of presence. Let us first start with the two separated, yet interrelated, mythemes in the second column: powerlessness and facelessness.

Table 1. Mytheme Breakdown of Slender Man Narratives.

Untouched	Left Behind	Taken
Normal	Deformed	Missing
The World	Transformed	The Outside
Healthy	Unhealthy	No Health
Friendly	Violent	Removed via Violence
Themselves	Act Unlike Themselves	Themselves Gone
	Powerlessness	
	Facelessness	

The Slender Man is most often depicted without a face, and so the concept of 'facelessness' seems an easy association to make with the Slender Man. However, the connections of lacking a face go much deeper. The faceless monster is not new with the Slender Man – he echoes the idea of cosmic fear in the work of H.P. Lovecraft (1973), where there is a grand fear in that which the human mind cannot comprehend. The faceless man is also found in surrealist art, where it symbolises the human condition (Bohn, 1991). For the Slender Man, the facelessness is not limited to the monster.

Characters in the narratives on the *Something Awful* forum (and in other Slender Man narratives outside of this context) are always in a stage of transition. In *Something Awful*, most characters began like everyone else – a normal member of society, untouched by the monstrous (see the first column above). However, these character's interactions with the Slender Man change them. They become set-apart from society. In Mr. 47's (2009) newspaper article, a young girl witnessed her sister being taken by the Slender Man, but she is overlooked. Similarly, in genesplicer's (2009) narratives, the children's account of the Slender Man in the woods is equally discounted. Their interactions with the monstrous remove them from an active part of a trustworthy society, though they are not as removed as those the Slender Man has taken, thus I have labelled them as 'left behind' (see the second column).

The Slender Man's narratives are communal act of storytelling. Many online narratives, particularly horror narratives, are told using 'in-character' communication. Essentially, users assert the truth and veracity of the narratives they share without ceding their position as author or audience member. This form of storytelling is best detailed in the horror sub-reddit No Sleep, wherein the rules for posting and commenting state that: 'everything is true here even if it isn't'. Those with narratives would claim they 'found' the documents, emails or photographs and are sharing these finds with the online community (see e.g. LeechCode5, 2009; TrenchMaul, 2009). Some are more performative in nature, including one from Victor Surge, where they posit an on-going experience (Victor Surge, 2009d), or that while reading the stories on the forum thread they can hear Slender Man's characteristic tapping at their window (KatWithHands, 2009; An observer 2009; Verloc, 2009). In an interview conducted as part of my fieldwork, one user commented that keeping to in-character communication 'keeps the story alive' while another explained how this causes the narrative to feel 'more real than novels or movies because of interactivity and being posted on message boards/places you could realistically believe someone might post their personal horror story'.

The interactivity of in-character communication encourages transformation of the text, whether by act of author or audience (Cover, 2006). The storytelling remains communal, despite individual creators arising and taking the narrative into a form they find personally most comfortable or evocative, leading to a truly transmedia story (Jenkins, 2006). This is echoed in the Slender Man's Tulpa Theory. Tulpas, originally derived from Theosophy (Mikles and Laycock, 2015), are thoughts which are given physical form. According to the Tulpa theory, the Slender Man may have been started on the forums of an online comedy website, but the consequence of many people thinking and contributing to the Slender Man resulted in him physically taking form as a Tulpa. This Tulpa theory is one discussed purely in play, but the concept underlying demonstrates the way in which individual creativity was allowed to flourish under one larger mythos. Each individual author can have their personal interpretation shine as canonical.

Despite this, in-character communication highlights the force the community holds above the individual. While the author may hold creative license, the authority of authorship is often non-existent within the context of in-character communication – the ultimate authority is given to the community rather than the individual. The originator of the Slender Man, Victor Surge, never corrected or claimed the Slender Man narrative as his own intellectual property (Slender Nation Podcast, 2011). The Slender Man is, and always has been, communal property (see Chess, 2012). Again, this is not unique to the Slender Man, but is representative of how mass communal storytelling works. In-character communication demonstrates the way in which facelessness is not only reserved for the Slender Man himself but is also put onto his audience.

The act of writing itself is a faceless one. The author's work is shared separate from author, becoming part of the communal narrative itself. This also echoes the experience of the characters in these narratives. They become faceless to society. Powerlessness is tied to facelessness. As both characters and audience become faceless, they lose the individual power they had – becoming less so until they are fully removed, or taken, from society.

This leads us to consider the three columns. These represent the three primary categories found across the narratives: those *untouched*, those *left behind* and those *taken*. 'Untouched,' used to distinguish it from the idea of the physical touch, which is important in relation to the ways characters shift from one character to another. 'Taken', are those who are missing, those who are, following the phrasing from Aleph Null (2009) on the Outside. The *taken* are fully removed from society; those the Slender Man has removed completely, making them disappear as if in thin air.

Those *left behind* are a different category from the other two, yet one derivative. These are those who have been forcibly removed from normal society through their experience of the Supernatural Other. Like the *untouched*, they are still members of society and present, but they are different. They have been altered and changed. Their roles as upstanding members of society is shifted, and their contributions are questioned. In most of the *Something Awful* forum narratives, the victims of this marginalisation are children. For example, in Mr. 47's (2009) newspaper narrative, the witness, a young girl named Alice, was treated with little attention, despite her testimony to the disappearance of her sister. Other narratives describe the children as being overly 'creative' with their thinking (genesplicer, 2009). When the distancing occurs to adults, they are often depicted as insane. Several of these accounts take shape in the form of documents or journals from asylums (see MonkeyMaker, 2009; The Black Nerd, 2009; Millard Fillmore, 2009; JMHX, 2009). In the web series *Marble Hornets* (2009–2014) the young adults the narrative is focussed on often have a denotation of mental health problems, including one character which features a 'reveal' of his many treatments (Entry 60.5). Despite not being taken by the Slender Man themselves, these characters (and sometimes locations) are still in some way removed from the society of the *Untouched*. While not as dead as those *taken*, they are *socially* dead, that is, in between the two categories while still holding traits of both. Their liminality is a form of social death due to the inability to remain *untouched* in society, and yet cannot be mourned as they are not gone.

The *left behind* is a structural mediator. Three types of relations occur with mediators in relation to the other categories. The first is a positive relationship, where there is movement between the two categories; something, or someone in the case of the Slender Man narrative, can move from one category to the other. The second is a negative relationship; negative relationships do not allow overlap or movement between categories. There is no ability for movement or transition between categories. The third type is a neutral relation where there is only some movement and overlap. Neutral relations are less likely to be found.

The three categories are interrelated by their movement. However, progression only occurs in one direction. While a character can start in society, as *Untouched*, once they move to a category in which their perspective on life has changed, they are unable to return to their previous category. One can move all the way through the categories to *Taken*, however once *Taken* they cannot return to life and society again. Essentially, this means the *Untouched* have a positive relationship with the *left behind*, and the *left behind* have a positive relationship with the *Taken*. However, the *Taken* have a negative relationship with the *Untouched*. The positive relationships are only in one direction. These relationships can be drawn as such:

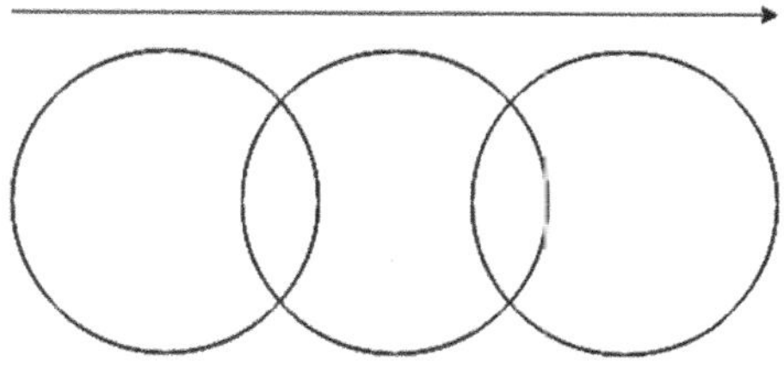

Fig. 1. Sketched Mytheme Relationships.

The overlapped circles depicted above (Fig. 1) demonstrate a positive mediation, with the above arrow illustrative of the sole path of progression in the structural model.

The primary consequence of a triadic model is the extension of the mediatory category. In our above understanding, the *Left Behind* is an extended position, a category in and of itself rather than an overlap between the two. The effect is an emphasis on the anxiety of the middle category. In order to better understand this, we have to reflect on an important aspect of the list of mythemes which appears to be missing: the Slender Man himself.

The Slender Man does not fit into any of the categories listed. He is not *Untouched* – he is not of a 'normal' society, which also means he cannot be *Left Behind*. He is also not *Taken*, as he is fully present. The categories detailed are based on interactions with him. Even though he is not a direct member of any category, he is directly responsible for each category.

In Jeffrey Cohen's (1996, 6) work, which set out the main dimensions of the interdisciplinary field of monster theory, he determined that one of the theories in which monsters function is as a determination of 'categorical crisis'. Anthropology has not made many strides to join the interdisciplinary monster theory (Musharbash and Presterudstuen, 2014), however structuralism can demonstrate this type of monster well. Mary Douglas's (2001) *Purity and Danger*, for example, contended that Israeli food laws led to a marginalisation of the pig as a monster of categorical crisis. By this understanding, the monstrous nature of the Slender Man is highlighted not only by his monstrous actions, but his fearsome inability to fit into the categorical understandings of the social formulation of three categories.

Combining the lack of categorical understanding with his fearsome nature to forcibly shift the categories of characters, the *Left Behind* category is one of anxiety. Characters are not free to move categories on their own, but are stuck in situ until the outside force of the Slender Man moves them to an alternate place. The characters are completely powerless to their own position. This powerlessness is also on the audience and writers who are forcibly inserted into the categories of the structure. This can be read against the previously considered Tulpa theory, demonstrating how even the individual creator is powerless in their position.

The extended mediatory category, therefore, demonstrates the extended anxiety felt within the category itself. Put another way, the mediatory stage is typically a temporary liminal stage, though here can be permanent. As a prolonged example, let us consider the typical dyadic structure, with mediatory overlap (i.e. would be pictured as two circles in a Venn Diagram), as categorical representation of a ritual. An initiate would begin in the first category – say the uninitiated society. After the ritual, they become part of a different segment – one which is apart and initiated. The ritual itself which leads to this change is a liminal position, where the individual is both initiated and uninitiated simultaneously. In the triadic model, this liminal position is extended. The individual could be trapped within the ritual of which their engaged, unable to return to *untouched* society in either stage. It is possible, in the triadic model, for the individual to live the rest of their life in the liminal stage. It is this entrapment that causes strife and anxiety for the individual caught in such a position.

The Digital Self

The triadic model of categories is not restricted to only the narratives themselves. This also presents an understanding of how the digital community of creepypasta sees society, and themselves, in the digital world. The term 'creepypasta' came from 'copy paste', a term used in the early stages of online communication when it was necessary to copy the words, and paste it elsewhere in order to share. The term for these shared narratives became 'copy pastas', and the horror variants 'creepypastas'. This particular section is derived from a more detailed look into the nature of 'digital culture' – the nature of the relationship between the digital self and digital community will be explored here.

Early studies of the internet and the online environment began to use the term 'digital culture', though this was separated from any definition or detailing of what 'digital culture' is comprised of (see Gere, 2002; Wiggins and Bowers, 2015). John Foley uses the term 'culture', but sees it as working differently than what is implied in the use of Gere or Wiggins and Bowers. He understands culture as a network, because they provide avenues for change. Culture, for Foley (2012, 68), thrives 'by remaining forever in play'. Milad Doueihi (2001, 12) also uses the term 'digital culture', which he defines as that which is 'comprised of communication and information exchange modes which displace, redefine, and reshape knowledge into new forms, formats and the methods for acquiring and transmitting this knowledge'. His abstract definition allows for the development of new and constantly changing forms of literacy, allowing the different forms of online communication to shift while not altering our understanding. This relates directly to Foley's (2012) concept that culture is never static, but rather constantly in play.

This definition may be preferable – if we are to define digital culture at all. The digital environment is constantly shifting, leading to a difficulty in pinning down any solid understanding. If there is to be a concrete understanding of digital culture, then it should be defined by the creativity of its users. The mythology of the Slender Man is a pivotal case demonstrating how storytelling can shape the online environment, brought together by the collective imagination.

I stress here the *if*'s present in the above paragraph. Roy Wagner's (2016) statement that culture, as anthropologists have learned to accept it, has been invented is of particular pertinence for the digital world. Through the process of cultural relativism, anthropologists have constructed the idea of culture, particularly culture as a thing in and of itself. Similarly, the term 'culture' has been used in relation to the digital to shift from a more generic use of the term to other considerations of social cohesion. I believe the notation of a complete 'digital culture' is neither relevant nor helpful.

Instead, the digital environment should be conceived of as many communities, rather than one culture. Healy (1997) has described how a community is spontaneous, emerging in constantly changing circumstances. The community aspect to the online environment is of primary concern for participants, and therefore for us as academics. The online environment at this current point is too diverging and wide-reaching in order to be consider as one cohesive culture, but rather a virtual geography within which multiple communities exist.

This is where the Slender Man, and his monstrous nature, becomes an interesting concern. The digital community which gave birth to the Slender Man is the creepypasta community. The ultimate goal of copypastas, and creepypastas in particular for our concern, is to be shared outside from authorial name. In essence, the ultimate aim of a creepypasta is to become an online urban legend. Whether or not this has already been fully achieved is a much larger debate than this chapter seeks to answer. The main point of the community, an important aspect to consider here, is to remain faceless.

The Slender Man's lack of existence in the structure demonstrates a key aspect of the monstrous presented by Jeffrey Cohen, that a monster demonstrates a 'categorical crisis' which disrupts the basic cultural categorisation (Cohen, 1996). This conception is repeated in many other analytical conceptions of the monster (Carroll, 1990; Gilmore, 2003; Asma, 2009; Mittman, 2013), and is what can be seen in the formation of the Slender Man. The Slender Man's lack of presence in the structure is simply because he does not fit within it. He disrupts the culture of his own community through his existence, and disrupts the life of the characters through his existence.

If we remove the overarching scary monster of the Slender Man with a different outside force, our previously detailed understanding of the world does not change drastically. The Slender Man provides a larger metaphor for the community to express their dissent with the issues of larger, typically 'western', social norms. The monster of the Slender Man is a physical expression of the social forces of the non-virtual outside world. Narratives demonstrate a commentary on the views of those with mental health issues in wider society: those who are marginalised and suffer because of it. Those who are 'Left Behind' are often considered as insane, resulting in their testimonies being silenced. The category is also equated with poor health, primarily in the *Marble Hornets* series (2009–2014).

Fieldwork conducted on a horror writing loci online exemplifies this contribution and social critique. Many discussed their connection to horror writing as a way of expressing themselves after childhood trauma. These parties saw the world in a worse, and more horrific, way than stories they read or saw expressed, and so they wrote their own.

The anxiety of the *Left Behind*, in many ways, expresses how the authors viewed their own digital self. If we change our understanding of the categories to reflect the various levels of physical to virtual, something interesting occurs. The *Untouched*, in this scenario, would be that of the physical world, the community more physically present around the user. The *Taken* would be the virtual world, and the digital community in which the user finds support, community, etc. As such, the user finds themselves in a strange middle ground between the two. In 2009 at the genesis of the Slender Man, the concept of an online community and a virtual home was not as readily found nor as commonly accepted. This is reflected in some early studies of the online environment, best summed up by Fialkova and Yelenvskaya's (2001, 87) study which stated that virtual communication and environment 'can only give an illusion of friendship, involvement, and belonging'. The concept of an online community is still not homogenous. Some of the language used online reflects the dichotomy between virtual and physical, including the designation of 'IRL', or 'in real life' – essentially leaving the virtual environment as the one which is not 'real'. This means that any substantial relationship or connection made online lies somewhere in between what is real and what is not real: a strange mediatory category in which their full online experience exists. This can provide a great sense of anxiety for the individual members of the online environment. They are *Left Behind*, controlled by a monstrous outside force.

Conclusion

In 2009, the internet gave birth to a monster. The sharing of images, stories, web videos, video games and other forms of digital content led to mass communal online story storytelling, with innumerable individuals contributing to the great online mythology of the Slender Man. Meanwhile, the monster works its dark works – its figure lingering in the back of photos a demonstration of its threat. Of how it steals children, and brutally kills its adult victims. The online community engaged with the myth both directly, in their telling and re-telling, and indirectly, by discussing the myth with a sincerity which, at first glance, appears intense. But this sincerity is in an act of play – users play with the stories, religious language and sincerity in their engagement of a mythology of death. The stories themselves play with the abstract boundary between the digital and the non-digital worlds. The structure of the full mythos – the collection of the various forms the narrative takes – demonstrates a horrific relationship of monster to society, but more importantly society to itself, which the myth also playfully places the audience into – so they become trapped in the play with death in the digital, which follows them even when the computer is off.

Interactions and Sensoria

Blue Isle Studios and Parsec Productions. 2015. *Slender: The Arrival*, Blue Isle Studios and Midnight City.

Cox, J. 2015. The scary game squad – Slender: the arrival [Playlist], *YouTube*, 26 January. Available at: http://bit.ly/2WiNT41

Kissel, B., Parks, M. and Zebrowski, H. n.d.. Ep. 323: Men in black part I: You fed the Tulpa. *Last Podcast on the Left*. Available at: http://bit.ly/32SRwjF

References

Aleph Null. 2009. *Create Paranormal Images*, Something Awful. Available at: http://bit.ly/31OlMup

An observer. 2009. *Create Paranormal Images*, Something Awful. Available at: http://bit.ly/2MTenWI

Asimos, V. 2019a. *The Slender Man Mythos: A Structuralist Analysis of an Online Mythology*. PhD thesis, Durham, Durham University Available at: http://etheses.dur.ac.uk/12962/

Asimos, V. 2019b. Video games as contemporary mythology, *Implicit Religion*, 21(1), 92–109. http://dx.doi.org/10.1558/imre.34691

Asma, S.T. 2009. *On Monsters: An Unnatural History of Our Worst Fears*, New York, NY, Oxford University Press.

Bascom, W. 1984. The forms of folklore: prose narratives. In *Sacred Narrative: Readings in the Theory of Myth*, Ed. A. Dundes, pp. 5–29, Berkeley, CA, University of California Press.

Blue Isle Studios and Parsec Productions. 2015. *Slender: The Arrival*, Blue Isle Studios and Midnight City.

Bohn, W. 1991. *Apollinaire and the Faceless Man: The Creation and Evolution of a Modern Motif*, Rutherford, Fairleigh Dickinson University Press.

Carroll, N. 1990. *The Philosophy of Horror, Or, Paradoxes of the Heart*, New York, NY, Routledge.
Chess, S. 2012. Open-sourcing horror: the slender man, marble hornets, and genre negotiations, *Information, Communication & Society*, 15(3), 374–393. https://doi.org/10.1080/1369118X.2011.642889
Cohen, J.J. 1996. Monster culture (seven theses). In *Monster Theory: Reading Culture*, Ed. J.J. Cohen, pp. 1–20, Minneapolis, MN, University of Minnesota Press.
Cover, R. 2006. Audience inter/active: Interactive media, narrative control and reconceiving audience history, *New Media & Society*, 8(1), 139–158. https://doi.org/10.1177/1461444806059922
Doueihi, M. 2011. *Digital Cultures*, Cambridge, MA, Harvard University Press.
Douglas, M. 2001. *Purity and Danger: An Analysis of the Concepts of Pollution and Taboo*, London, Routledge.
Fialkova, L. and Yelenevskaya, M. 2001. Ghosts in the Cyber World: an analysis of folklore sites on the Internet. Fabula 42, 64–89.
Foley, J.M. 2012. *Oral Tradition and the Internet: Pathways of the Mind*, Urbana, IL, University of Illinois Press.
Frazer, J.G. 1994. *The Golden Bough: A Study in Magic and Religion*, Oxford, Oxford University Press.
Geertz, C. 1973. *The Interpretation of Cultures: Selected Essays*, New York, NY, Basic Books.
genesplicer. 2009. *Create Paranormal Images*, Something Awful. Available at: http://bit.ly/2otxGfS
Gere, C. 2002. *Digital Culture*, London, Reaktion Books.
Gilmore, D.D. 2003. *Monsters: Evil Beings, Mythical Beasts, and all Manner of Imaginary Terrors*, Philadelphia, PA, University of Pennsylvania Press.
Hamayon, R. 2016. *Why We Play: An Anthropological Study*, Translated by D. Simon, Chicago, IL, Hau Books.
Healy, D. 1997. Cyberspace and place: the internet as middle landscape on the electronic frontier. In *Internet Culture*, Ed. D. Porter, pp. 55–68, New York, NY, Routledge.
Jenkins, H. (2006). *Convergence Culture: Where Old and New Media Collide*, New York, NY, New York University Press.
JMHX. 2009. *Create Paranormal Images*, Something Awful. Available at: http://bit.ly/2MTEWuM
KatWithHands. 2009. *Create Paranormal Images*, Something Awful. Available at: http://bit.ly/2MRHUjg
Kunin, S.D. 2003. *Religion: The Modern Theories*, Edinburgh, Edinburgh University Press.
LeechCode5. 2009. *Create Paranormal Images*, Something Awful. Available at: http://bit.ly/2MW1ETb
Lincoln, B. (1999). *Theorizing Myth: Narrative, Ideology, and Scholarship*, Chicago, IL, University of Chicago Press.
Lovecraft, H.P. 1973. *Supernatural Horror in Literature*, New York, NY, Dover Publications.
Luhrmann, T.M. 1989. *Persuasions of the Witch's Craft: Ritual Magic and Witchcraft in Present-day England*, Oxford, Blackwell.
Malaby, T.M. 2009. Anthropology and play: The contours of playful experience, *New Literary History*, 40(1), 205–218.
Malinowski, B. 1948. Myth in primitive psychology. In *Magic, Science and Religion and Other Essays*, Ed. B. Malinowski, pp. 72–124, Glencoe, The Free Press.
Marble Hornets. 2009–2014. Marble hornets [Videos]. *YouTube*. Available at: http://bit.ly/2pY5YYS
Mikles, N.L. and Laycock, J.P. 2015. Tracking the Tulpa: Exploring the "Tibetan" origins of a contemporary paranormal idea, *Nova Religio: The Journal of Alternative and Emergent Religions*, 19(1), 87–97. https://doi.org/10.1525/nr.2015.19.1.87

Miles-Watson, J. and Asimos, V. Eds 2019. *Bloomsbury Reader in the Study of Myth*, London, Bloomsbury.

Millard Fillmore. 2009. *Create Paranormal Images*, Something Awful. Available at: http://bit.ly/32WofnW

Mittman, A.S. 2013. Introduction: the impact of monsters and monster studies. In *The Ashgate Research Companion to Monsters and the Monstrous*, Eds A.S. Mittman and P. Dendle, pp. 1–16, Surrey, Ashgate.

MonkeyMaker. 2019. *Create Paranormal Images*, Something Awful. Available at: http://bit.ly/2WlHlS6

Mr. 47. 2009. *Create Paranormal Images*, Something Awful. Available at: http://bit.ly/2BNXC95

Musharbash, Y. and Presterudstuen, G.N. Eds 2014. *Monster Anthropology in Australasia and Beyond*, New York, NY, Palgrave Macmillan.

Ortner, S.B. 1999. *Life and Death on Mt. Everest: Sherpas and Himalayan Mountaineering*, Princeton, NJ, Princeton University Press.

Parsec Productions. 2012. *Slender: The Eight Pages*, Rio Rancho, NM, Parsec Productions.

Raj, S.J. and Dempsey, C. 2010. Introduction: Ritual levity in South Asian Traditions. In *Sacred Play: Ritual Levity and Humor in South Asian Religions*, Eds S.J. Raj and C. Dempsey, pp. 1–20, Albany, NY, State University of New York Press.

Slender Nation Podcast. 2011. *VICTOR @#$%ING SURGE* [Podcast episode].

The Black Nerd. 2009. *Create Paranormal Images*, Something Awful. Available at: http://bit.ly/2Wk0WBZ

TheRiffie. 2009. *Create Paranormal Images*, Something Awful. Available at: http://bit.ly/2MQS18c

TrenchMaul. 2009. *Create Paranormal Images*, Something Awful. Available at: http://bit.ly/2BNxsmJ

Tylor, E.B. 2010. *Primitive Culture: Researches into the Development of Mythology, Philosophy, Religion, Art, and Custom*, Cambridge, Cambridge University Press.

Verloc. 2009. *Create Paranormal Images*, Something Awful. Available at: http://bit.ly/31XtSBj

Victor Surge. 2009a. *Create Paranormal Images*, Something Awful. Available at: http://bit.ly/2pVJKqu

Victor Surge. 2009b. *Create Paranormal Images*, Something Awful. Available at: http://bit.ly/2MW1ETb

Victor Surge. 2009c. *Create Paranormal Images*, Something Awful. Available at: http://bit.ly/2BNxsmJ

Victor Surge. 2009d. *Create Paranormal Images*, Something Awful. Available at: http://bit.ly/2BNxsmJ

Wagner, R. 2016. *The Invention of Culture*, 2nd ed., Chicago, IL, The University of Chicago Press.

Wiggins, B.E. and Bowers, G.B. 2015. Memes as genre: a structuralist analysis of the memescape, *New Media & Society*, 17(11), 1–21. https://doi.org/10.1177%2F1461444814535194

Chapter 10

Gameful Interactions: The 'Ludification' of Zombie Fiction

Chloé Germaine Buckley

Abstract

Cultural perceptions of the zombie have shifted dramatically in the twenty-first century. No longer only associated with anxiety and fear, zombie fiction often appeals to pleasure. One source of pleasure comes from ludification, the process whereby game-like principals and gameful elements shape non-game activities. Increasingly, print fiction borrows from games and uses ludic elements to shape narratives. As such, it has become embedded in convergence culture, a dynamic media ecology where top down processes compete with bottom up processes. This chapter argues that ludified zombie fiction brings this media ecology into sharp relief, revealing ways that gamification and ludification are just as apt to reinforce capitalist processes of commodification and neo-liberal ideologies of power as they are to dismantle them. Through a close reading of three contemporary zombie fictions, this chapter exposes tensions and contradictions in ludification. The dead body of the zombie, the nihilistic landscape of the post-zombie apocalypse and the futility of human endeavour in the face of walking death are all elements of genre that undercut the gamified pursuit of external utility-oriented goals. The chapter explores these knotty ethical and ideological problems, not only considering the zombie apocalypse as a gameful space for rethinking social organisation, but also recognising it as a platform for the promotion of neo-liberal ideologies that perpetuate existing power inequalities through coercive disciplinary regimes.

Keywords: Zombie fiction; contemporary literature; ludification; gamification; convergence culture; ethics

Death, Culture & Leisure: Playing Dead, 139–154

doi:10.1108/978-1-83909-037-020201017

Introduction: Gamification, Ludification and Convergence Culture

In the twenty-first century, the figure of the zombie has become increasingly playful. One manifestation of the playful zombie emerges through 'gamification', a process whereby game-like design principals and 'gameful' elements, which aim to elicit experiences common to gameplay, shape non-game activities (Koivisto, 2017, 17; Deterding et al., 2011). Gamification points to a dynamic process of adaptation. Not only have zombies been adapted into game formats, but also game formats increasingly shape other media. In print fiction, ludic elements create interactive narratives and contribute to transmedia franchises, where the book is one component of a larger game world. This chapter examines examples of gamification in zombie fiction from Max Brooks' *The Zombie Survival Guide* (2004), through Max Brailler's 'choose-your-own-adventure' novel, *Can You Survive the Zombie Apocalypse?* (2011), to Naomi Alderman's *Keeping Fit and Living Well in the Current Zombie Emergency* (2016), a companion to the fitness app, *Zombies, Run!* (2012) that Alderman co-created with developer Six to Start. I tease out ambivalences that inhere in gamified zombie fiction, arguing that it encapsulates tensions at the heart of intersecting twenty-first-century cultural modes: the 'ludification of culture' defined by Joost Raessens (2006) and 'convergence culture' theorised by Henry Jenkins (2006a).

The term 'gamification' originated in entrepreneurship and marketing to describe the exploitation of game design principles to stimulate consumer engagement. However, Fuchs, et al. (2014, 7) argue that gamification should be understood beyond this context as an essential component of society and culture. Likewise, Raessens (2014, 95) identifies gamification as one aspect of the 'ludification of culture'. Raessens (2006) argues that ludification in the twenty-first-century arises through the effect of computer games and other digital technologies in stimulating playful goals and identities. Zombies exemplify Raessens' argument and demonstrate the wider contexts of gamification. From early examples such as Atari's *Entombed* (1982), zombie-themed computer games have proliferated, typifying the survival horror genre with enduring classics such as *Resident Evil* (Capcom, 1996). The popularity of zombie games has contributed to the monster's cultural resurgence, seen in the success of television shows like *The Walking Dead* (AMC, 2010–Present). In 2015, the *BBC* capitalised on the popularity of the zombie with its television game show, *I Survived a Zombie Apocalypse*, which asked contestants to participate in a live-action roleplaying game that used scenarios familiar from zombie films and video games. These interconnections suggest that the process whereby zombies are bound up in the ludification of culture involves the hybridisation of games with other forms, including film, television and print fiction. The apotheosis of this development is the fitness app, *Zombies, Run!* (Six to Start and Alderman, 2012), which uses zombies to promote user engagement with a lifestyle product.

Beyond games, twenty-first-century zombie texts borrow or 'cannibalise' across forms to create transmedia stories and hybrid narratives. Alderman's guide, for example, is a book based on an app, but there are also games based on comics-turned-television-series such as Cryptozoic's *The Walking Dead* board game (Jones, 2011), and films based on video games such as the *Resident Evil* franchise (2002–2016). The disruption of traditional patterns of adaptation reveals that zombie fiction is embedded in what Jenkins (2006a, 2) calls convergence culture: a paradigm shift designating the flow of content across multiple media platforms, the migratory behaviour of media audiences and the complex interactions between grassroots and corporate power. Old and new media forms interact in complex ways. Not only is the direction of adaptation inverted, but also top down processes (determined by producers and companies) compete with bottom up processes (the demands of consumers). Convergence culture is thus riven with tensions that manifest within and without zombie fiction. In this dynamic media ecology, zombie fandom is a substantial collective presence with the power to shape media content. Similarly, the zombie apocalypse narrative demands a collective response and the sharing of knowledge and resources that exist outside institutions and authoritative structures. However, collaborative social models compete with a neo-liberal individualist ideology that shapes gameful engagements. Though the zombie apocalypse may seem an ill-fitting backdrop for corporate commodification, *Zombies, Run!* fans are encouraged to subscribe to extra content and don their 'Runner Five!' T-Shirts whilst exercising.

Just as convergence culture involves competing forces, so, too, do 'ludification' and 'gamification'. Fuchs et al. (2014, 9) note that the latter term has too easily been assimilated into the vocabulary of designers and critics such that it requires further interrogation. They are wary of a claim made by designer Jane McGonigal (2011) that gamification promotes participatory practices and that it might transform the world. Raessens (2014, 93) also cautions against utopian interpretations of ludification since 'play is often less open than it looks'. He suggests that not all examples of ludic media are 'harmless disinterested fun' and might easily become implicated in political ends (Raessens, 2014, 104). Ian Bogost (2011b) goes further, rejecting gamification as 'a grifter's game'. This dichotomous response to gamification depends on the critics' perspective on convergence culture: if gamification is deemed to be in the control of a participatory community of users then it is all for the good (McGonigal, 2011); if it is a tool that benefits greedy companies wanting to sell more stuff to gullible consumers then it is to be resisted (Bogost, 2011a, 2011b). Zombie fiction brings this media ecology into sharp relief, revealing that gamification and ludification are just as apt to reinforce capitalist processes of commodification and neo-liberal ideologies of power as they are to dismantle them.

From Social Anxiety to Gameful Interactions: Playful Zombies?

Like debates about the merits of gamification, criticism and commentary on the zombie are also interested in questions of economic and social power. The most common manoeuvre positions the undead as a figuration of social anxieties or as a metaphor for the loss of identity and agency in uncertain times (Platts, 2013; Tenga and Zimmerman, 2013; Abbott, 2016). Though negative associations of the zombie persist in critical works, alternative readings are beginning to emerge that account for the pleasures offered by some zombie texts. Kyle William Bishop (2015), for example, considers the appeal of sympathetic zombies as a form of relief from present-day fears and anxieties. Beyond eliciting sympathy, zombies might also facilitate narratives of agency and resistance. Emma Austin (2015) argues the latter in her reading of the 'zombie walk' as a 'carnivalesque' practice able to destabilise social hierarchies. Again, Austin's reading is concerned with power relations. Yet, the zombie is as much caught up in the tensions and complexities of convergence culture as are games, which complicate any radical potential. Sarah Juliet Lauro (2011, 227–228) suggests that zombie fan culture, which includes grassroots events like the zombie walk, is vulnerable to appropriation by corporate power.

Playful zombies have also been taken up by Game Studies, which has begun to focus on the monster as a ludic antagonist. However, here also there remains an emphasis on anxiety. Backe and Aarseth's (2014, 1–2) examination of zombieism in video games opens with an implicit endorsement of a social anxiety reading, evoking 9/11 as the source of the zombie renaissance in games. They examine Tanya Krzywinska's (2008, 153) assertion that the popularity of the zombie in video games depends on their status as the ideal enemy without soul or consciousness. Backe and Aarseth problematise this reading, unhappy with the ethical dimension of games in which zombies appear as a target for violence. Positing the zombie as an allegorical figure, they evaluate a series of games, denigrating those that ignore the history of zombie iconography and foreclose critical engagement and praising games that facilitate players' engagement with the political dimension of the allegory. Some of the elements Backe and Aarseth praise feature in the print fiction texts discussed in this chapter. The use of the zombie as faceless target for violence is part of the appeal of Brooks' *Survival Guide,* but so, too, is his ironic reflexivity about the history of zombie iconography, which in turn suggests that the reader might occupy a critical position. Gamified fictions such as *The Zombie Survival Guide* reveal the tensions and ambiguities in gameful uses of the zombie that are elided by Backe and Aarseth's (2014, 6 and 13) insistence on the zombie as a negative allegory.

My analysis builds on Backe and Aarseth's questions, widening the scope to examine non-game texts. In so doing, I chart a history of zombie narratives that sheds light on trends of ludification and gamification in convergence culture outside of games themselves. Brooks' *Zombie Survival Guide* (2004) is an early example of convergence culture, borrowing from film and video games in its creation of a secondary world that imagines an imminent zombie outbreak. Brooks' book is ironic and gameful, eliciting reader participation in the fantasy with a direct address through the medium of the 'guidebook'. This narrative voice is developed further in the gamebook format adopted by Max Brailler in *Can You Survive the Zombie Apocalypse*? (2011). Also borrowing tropes from film and video game, Brailler resurrects the 'choose-your-own-adventure' gamebook, codifying interactivity at the level of narrative structure in an explicit ludification of the print format. Part of the pleasure of Brailler's text is in its nostalgic appeal; it harks back to analogue game forms from the 1970s and 1980s that prefigure the video game.

Indeed, this book forms part of a wider resurgence in the adventure gamebook, signalled by the publication of *Blood of the Zombies* by Ian Livingstone in 2012 to mark the thirtieth anniversary of the *Fighting Fantasy* series. In the same year, Naomi Alderman developed the lifestyle app, *Zombies, Run!* The app exemplifies gamification because the undead are deployed to encourage users to pursue utility-oriented goals external to the game world: a couch-to-5k fitness programme. The app quickly developed a transmedia story world, and the alternative post-zombie-apocalypse landscape inhabited by 'Runner 5' was given further 'consistency' in Alderman's (2016) companion book, which builds on Brooks' survival guide, not only offering advice on how to survive the 'zombie emergency' but how to live 'well'. Albeit in different ways, these texts reveal that gamification and ludification are enfolded in convergence culture. These are hybrid narratives that develop gameful engagements but this is not without its contradictions. Indeed, the zombie is an apt figure to draw out the tensions of convergence culture. As the zombie becomes an increasingly playful figure, it demands complex critical engagements with its disruptive presence.

Ironic Survival(ist) Horror

Though it predates the rise of gamification, Max Brooks' *Zombie Survival Guide* (2004) hybridises literature with ludic elements. In so doing Brooks inaugurates an explicitly playful sub-genre of zombie fiction, anticipating 'mash-ups' such as Seth Grahame-Smith's *Pride and Prejudice and Zombies* (2009) as well as more explicitly gamified works. Eschewing a traditional narrative, Brooks employs the guidebook format to construct a secondary world in which a zombie outbreak is imminent. The book offers advice on topics such as different types of zombies, weapons and combat techniques, home defence strategies, how to effectively hunt zombies, and long-term survival. At the end there is a list of recorded zombie attacks dating from 60,000 BC to the present, a gesture that makes use of the 'secret history' trope to lend consistency to Brooks' world-building and suggest that readers have always been living in the shadow of a zombie outbreak. Finally, Brooks provides an 'outbreak journal' for readers to complete with their own observations and preparations.

How far readers participate in Brooks' pretence is beyond the scope of this analysis, but the book invites immersion in the world of zombie survival beyond the usual parameters of fiction's suspension of disbelief: a gameful engagement. As Frans Mäyrä (2008, 109–110) notes, games invite different types of immersion through the interplay of story, rules, the spatial dimensions of the game and the affordances of the audio-visual technology. Evoking a secondary world in which the dead may rise at any moment, *The Zombie Survival Guide* invites immersion along 'narrativist' or 'dramatist' lines (Mäyrä, 2008, 82). The non-fiction format simultaneously lends this fantasy a dose of reality, but also playfully punctures that reality with self-reflexive gestures. For example, Brooks (2004, xiv) insists that the reader 'not discount any section of this book as hypothetical drama'. Such seriousness contrasts with Brooks' (2004, 22) playful denouncement of movie directors as the 'greatest enemy' of the zombie despite his acknowledged debt to the 'silver screen'. At the same time, Brooks also suggests a 'gamist' attitude to the undead threat. 'Gamist' immersion contrast with narrativist or dramatist modes because it suggests players will focus on overcoming game challenges and finding optimal strategies rather than becoming involved in the story (Mäyrä, 2008, 82). Brooks' (2004, 30) advice on weapons and becoming 'battle-ready', for example, evokes the gamist immersion of first-person-shooter video games. Much of his advice focusses on optimal strategies for the field of battle. The overriding intra-story-world goal is that of mastery: of the available knowledge about zombies, of physical disciplines like hand-to-hand combat and, finally, 'unwavering, unquestionable discipline' of self and others (Brooks, 2004, 125).

Brooks' promotion of mastery, capacity-building and discipline is also ideological. Abbott (2016, 6) documents Brooks' avowed inspiration in the 'culture of paranoia and survivalism' of the late-twentieth-century United States, which included fears of nuclear war and anxieties about natural disasters. These fears manifest in Brooks' insistence on martial discipline and violent masculinity and in violent encounters where you successfully stab a zombie through the eye with a bayonet or smash its skull with the 'wooden stock of a well-made, expertly used bolt- or lever-action rifle' (Brooks, 2004, 45). A tacit endorsement of the right to bear arms emerges when Brooks (2004) notes that civilian gun shops outstrip military armouries in the United States, a fact that will prove 'critical in any of the scenarios covered in the latter part of this manual'. Brooks (2004, xiv) also expresses libertarian sympathies in his dismissal of state organisations: This is a book for 'private citizens' who 'refuse to be victimised'. Indeed, Brooks delights in using the zombie to imagine the destruction of societal and governmental institutions. Woe betide the foolish citizen who flocks to the local police station in search of aid. The promotion of a gamist attitude emphasises these ideologies and adds elements of neo-liberal individualism. The mantra is 'train constantly': only those who obey a strict diet and physical regimen will successfully battle the living dead (Brooks, 2004, 29). Such conservative attitudes focus on the capacities of the individual, locating power in bodily capital and willpower. The scenarios that follow assume male, able-bodied, white survivors who possess the necessary economic independence to invest in extensive preparations.

Brooks' endorsement of individualism vies with communal models of social organisation demanded by his ludified zombie world, functioning as a microcosm for the tensions of convergence culture. Advocating the building of a 'fortress' on the one hand, Brooks (2004, 87 and 125) also suggests that a 'collective response' is required. In this respect, the survival guide exemplifies Jenkins (2006a) description of the 'knowledge communities' forged in a participatory culture. These emerge 'in realms where no traditional expertise exists' and create spaces in which 'the pursuit of and assessment of knowledge is at once communal and adversarial' (Jenkins, 2006a, 20). Brooks assembles a wealth of information from a variety of sources (official and unofficial) but scoffs at those who would ignore his sage advice, presenting himself as an authoritative elder. He seems to advocate what Jenkins (2006a, 4) calls 'Collective Intelligence': 'None of us can know everything; each of us knows something; and we can put the pieces together if we pool our resources and combine our skills'. However, the interlocutor is tasked with leadership and discipline (Brooks, 2004, 91 and 159). Brooks' notion of the group may rely on collective intelligence, but this is not a horizontal community, nor is it open: it depends on authoritarian leadership, isolation and the repulsion of rivals. This is far from the utopian culture described by Jenkins. Recently, Jenkins (2019) defines participation as the 'redistribution of power' and argues that it is a necessary component in tackling the global crisis in democracy. Yet, this very same crisis of democracy threatens the end of an emergent participatory culture. The tensions in Brooks' narrative reveal the challenges faced by social theorists like Jenkins. One the one hand, gameful interactions in the context of an imagined zombie outbreak suggest the importance of informal knowledge communities, of collaboration, and playful problem-solving. On the other hand, the empowering potential of these practices are subsumed by individualist ideologies promoted by the zombie game world and a gamist attitude.

The figure of the zombie further complicates this reading by dint of their status as an overwhelming threat. Gamist responses and goal-oriented interactions find resistance in the apocalyptic doom that pervades the story-world. Brooks (2004, 104) revels in the prospect of an apocalypse, describing numerous scenarios that end in death and defeat: 'more than one vehicle has been found alone in the wilderness, tank dry, blood-smeared, cabin empty', he tells you in a passage about preparing for long-distance travel. Brooks offers disasters without solutions, taking you down a path out of which there is no chance of escape as though you are reading a 'choose-you-own-adventure' without the opportunity to change direction. Perhaps you may escape a zombie incursion by jumping from one roof to the next, Brooks (2004, 117) muses, but what then? In such passages, the gamist approach is undercut. As it progresses, the text continually increases the scope of its sinister secondary world, eventually opening into an apocalyptic scenario:

> Let's assume that you have followed all these instructions correctly. The battle is over, the area is secure, the victims have been mourned, the zombies have been burned. Hopefully this will be the last time you will ever have to raise your hand to undead flesh. But what if it isn't? What if your struggle was merely one small theater of a greater, all-out war between the living and the dead? What if, heaven forbid, it is a war humanity loses? (Brooks, 2004, 153)

In the 'devastated landscape' of an undead world there is little opportunity for empowerment and Brooks' evident pleasure in exploring this worst-case scenario does not fit easily with the utility-oriented goals of gamification.

Agency and You: Reader as Player?

If narratives appear to endow their characters with agency, then ludic narratives that locate their reader as a character within a story world may seem to amplify this agentic property. However, as the reading of Brooks' text suggests, the zombie genre complicates questions of power and agency in numerous ways, not least by imagining the reader as one type of survivor. I suggest that the ludic structure of zombie narratives provokes further questions about the concept of agency, exemplified by Brailler's (2011) choose-your-own-adventure, *Can You Survive The Zombie Apocalypse?*. The title is an invitation to participate, and further questions are posed in the blurb that emphasise the 'choices' demanded: 'Moral dilemmas. Strategic decisions. Weapons. Vehicles. Will you be a hero? Or will you cover your own ass at all costs?' (Brailler, 2011). In this overture, Brailler suggests a gamist immersion whereby you can strategise to navigate the narrative's decision tree successfully. In addition, he offers narrativist immersion through 'moral' qualms and the necessity of having to 'withstand' the gruelling hours, days, and weeks of a long-term zombie outbreak.

The ludic structure also creates a gap between the player-character and the reader, which did not emerge in Brooks' guidebook format. 'You' are a young, white male, living alone in New York when the zombies attack. You exhibit character traits of a typical slacker, drinking beer and playing video games as New York descends into chaos, although there are opportunities for you to be brave, ingenious and even heroic, depending on what choices the reader makes. The macho and heterosexist elements of the text are sometimes distasteful, but enough critical distance emerges that such moments may feel like pastiche. The decision tree and different pathways, for example, create dialogism as the narrator explicitly judges your behaviour, from berating you for turning down the advances of the 'hottest girl in high school', to calling you an 'asshole' for abandoning a bus full of kids (Brailler, 2011, 329 and 299). In such moments the ironic distance between the reader, the character you inhabit, and the narrator is greatest, allowing space for reflection on some of the hackneyed tropes. Nonetheless, Brailler does nothing to challenge the figuration of agency as male, white and able-bodied.

The second-person address of the adventure gamebook simultaneously collapses and delineates the player and the reader. 'You' are both the hapless 'jackass' trying to survive and a reader who is somewhat distanced from the action. Indeed, the reader might even come into conflict with the responses of the player whose role they must inhabit. This is especially likely in Brailler's book. Beyond the potential for disjunction in terms of identification, the second-person address disrupts straightforward notions of immersion. Paul Wake (2016, 192) argues that the gamebook is thus paradoxical in nature because it 'situates the reader in multiple positions at once; here both dead and not dead, both character and reader'. In *Can You Survive?*, for example, one ending wraps up 'nearly two years later, when … the Army finds your body, frozen to the core' (Brailler, 2011, 260). This is an apt example of what Wake (2016, 193) identifies as an estranging form of address, unique to the adventure gamebook, which forges 'identification-through-dislocation'. The estrangement emerges in the simultaneous conflation and disjunction of 'the horizontal' world of the story and a 'vertical address' to a reader located outside (Wake, 2016, 193), which cannot be accounted for by theories of immersion applied to video games.

Whether or not the reader identifies with the player-character, he remains a largely ineffective 'agent' in the story world. The trope of the juvenile slacker determines that many of the pathways end in failure: you decide to go to the bar instead of escaping the city, for example. Beyond the demands of genre, the ludic structure overwhelmingly emphasises failure rather than mastery or escape. This means that, as in Brooks' work, the word 'survive' seems ironic. Over thirty of the fifty potential endings describe a gory death for the protagonist; half of these involve succumbing to the zombies. A further ten endings leave you stranded with no prospect of survival. That most paths lead to a violent death complicates the process whereby agency is seemingly handed to both player and character via the direct address. Indeed, the combination of the gamebook form and zombie genre suggests that conceptions of agency via reader participation or narrative interactivity are flawed. The classic account of agency in interactive fiction is Janet Murray's (1997, 16) description of 'the satisfying power to take meaningful action and see the results of our decisions and choices'. Although Brailler allows the reader to see in gory detail the result of each decision made, the actions undertaken by the player character are not meaningful. Elsewhere, agency is conceptualised through a psychological framework. Mary Ann Buckles (1985, 2) links agency in games to the psychological concept of 'effectance' – 'the desire for competence and feeling effective in dealing with the surrounding environment'. Again, this concept of agency is undercut by Brailler's zombie-themed gamebook, since in many cases you make limited or limiting choices that you know to be ineffective. Often you are given no choices at all, but told, for example, 'you sink into the seat, out of breath. Out of energy. Out of options. Turn to page 156' (Brailler, 2011, 150). These fatalistic moments accord with Esther MacCallum-Stewart and Justin Parsler's (2007, 14) concept of 'illusory agency' in games (2007, 14). They argue that what appear to be meaningful choices are simple binaries or non-choices limited by the game narrative. Brailler's binaristic or non-choices, then, eschew the notion of gamefulness as mastery or effectance, a fact emphasised by the generic demands of the zombie outbreak.

If the player character struggles to be agentic in a zombie game world what of the reader's position as an agent? In their survey of agency in games, Wardrip-Fruin et al. (2009) suggest it is best conceptualised as phenomenon involving player and game. This concept of agency echoes trends in philosophy that locate agency not as a subjective property but distributed across assemblages of human and non-human actors. In narratives, the human character – 'You' – is not an agent, but a figuration to which the narrative seems to ascribe agency. Close reading often reveals this agency to be an illusion, since it does not inhere in the subject but only emerges in coalition with external forces. However, when the game world is populated by zombies, elements of a crumbling infrastructure, desperate and fleeing citizens, and other environmental hazards, any agency that emerges in the interaction between player and game is highly fleeting as choices and intentions are swiftly derailed by other actors. That zombies are one such actor counters Krzywinska's explanation for their popularity in games and complicates a straightforward reading of the zombie as a metaphor for fears about the loss of agency. Indeed, the ironic distance between the reader and the player character casts doubt on any notion of agency that locates it as a subjective property, since you are free to lead your character to a gory, violent end. That is, the agency of the reader undercuts the idea that agency inheres in a character, especially if the estrangement between the two leads to a misalignment of identification or empathy. At the same time, the reader is constrained by the ludic structure. This is most obvious when the author pours scorn on the character for a poor choice, or else provides you with two equally poor choices. In such moments, reader agency dissolves along with character agency.

The ludic narrative also limits the scope in which action may take place. Wake (2016, 194) suggests that the narrative focalisation of the game book limits the reader to perceptible objects. This is very unlike the worlds-eye view provided by Brooks and results in a much more claustrophobic story world and game space. The limited perspective hampers the ability of the reader (in role as the player-character) to make meaningful choices, though genre literacy could mitigate these limits to a certain extent. However much meta-textual knowledge the reader possesses, though, they still find that story branches converge on gory ends, suggesting that the narrative is a labyrinthine trap from which there is no escape. Only five of the endings result in your character leaving New York. The other forty five end in some form of defeat, inviting you to return to the beginning, cycling back into the game world again and again.

Utility-Oriented Fiction: From App to Book

Zombies, Run! contrasts with Brooks' and Brailler's work because it is premised on the offer of success. Designed as a running app to help individuals complete a 'couch-to-5k' programme, *Zombies, Run!* appeared on android in 2012. It represents an explicit gamification of zombie fiction, deploying the undead as a motivating factor to promote personal development and fitness goals. *Zombies, Run!* also exemplifies transmedia storytelling and convergence culture since the narrative and game world extends across media platforms. The book is a later addition, masquerading as a manual issued by the intra-story entity, the 'Ministry of Recovery' to offer further advice for *Keeping Fit and Living Well in the Current Zombie Emergency*. Where my discussion of Brooks and Brailler's work was centred on the gameful interactions they invited within a story world, here I look at how Alderman develops an intra-story world in reference to an external world. The fictional narrative set in during a national zombie emergency explicitly promotes health and fitness goals, along with a lifestyle, grounded in real-world contexts. This exchange between fictional and external worlds amplifies the tensions of ludified zombies.

Firstly, zombies seem to provide a means to produce intrinsic motivation in the player/reader, a psychological state linked to the success of gamification. Koivisto (2017, 27) notes that psychological theories of self-determination suggest that competence, autonomy and relatedness increase the chances of experiencing intrinsic motivation, which in turn is linked to enjoyment and increased engagement in an activity. She further argues that these psychological theories have led to the justification of gamification as a positive experience that can contribute to well-being. Alderman's tone throughout the guide echoes this rhetoric of well-being, for example, when she suggests the reader starts their 'day off right' with a gentle stretching routine, or when she advises you to work your training around a 'challenging yet reasonable distance' (Alderman, 2016, 33 and 74). The book gives you blank pages on which to record your progress, allowing you to date and quantify the number of squat jumps you manage, or track your running distance and pace over time. These elements of the book are addressed vertically, outside the story world, which in turn functions as set-dressing for the serious business of fitness training. Yet, the story world is also more than this, as 'Mr G-V of Tottenham' tells the reader:

> Before the apocalypse, I hadn't paid much attention to exercise, to be honest ... I used to collapse on the sofa every night. But after the fall of London, I was stuck in my third-floor maisonette for eight weeks – I thought I might be there forever! The only thing that kept me sane was my workout routine. I started by just trying out some push-ups and walking on the spot. It was a real buzz to notice my muscles getting stronger, and it took my mind off the undead bodies of my decomposing neighbours growling outside. (Alderman, 2016, 17)

This case study reveals how Alderman employs the zombie narrative as a motivational element in service of utility-oriented goals, but, equally, how those goals feed back into the creation of her story world. Here, we learn of the fall of London through the narration of a character becoming trapped in their home, which is also a vehicle through which Alderman can advise on a fitness routine. Alderman also contrasts the rotting bodies of the dead with a healthy body, the constant presence and reminder of death functioning as a spur for the reader/player to celebrate and strengthen their own lively physical energies. Indeed, Alderman (2016, 32) insists that every stretch and bend we do reasserts our own humanity.

These shifts between the story world and external reality result in an ironic address that is at turns playful (you are not really practising squats so you can run from a zombie horde) and serious (but you must nonetheless remember to seek advice from your GP before undertaking exercise). The tensions cut both ways. Explicit utility-oriented goals interrupt the development of narrativist immersion, but narrative tropes likewise undercut the promotion of real-world goals. Some elements of the book do not serve external goals at all. For example, the reader is invited to jot down a to-do list to help them reinforce their home from zombie attack. Here, the zombie is not a handmaiden to gamification. Rather, gamified features give consistency to a fictional world with its own zombie history and mythology. The instability of gamification in *Zombies, Run!* is amplified by shifts in tone. Cheesy motivational copy ('whatever you do has value!') grates against shocking 'facts', including that 90% of the UK population has been killed, and gory descriptions of decomposing neighbours (Alderman, 2016, 14, 9, and 12). The result often borders on parody, creating ironic distance that allows Alderman to comment on the process of gamification. Indeed, she signals the disparity in the different types of language and draws attention to the gap between the game world and the real world when she assures readers that though it 'may seem disrespectful to turn an event as appalling as the zombie apocalypse into a game … the results speak for themselves' (Alderman, 2016, 95). Her commentary also endorses theories of gamification linked to positive motivation, relatedness and engagement: 'The critical thing is not the nature of the project but that you should feel engaged in it and be able to see utility in your own life' (Alderman, 2016, 40). Yet, it is difficult to see the utility of, for example, pictogram instructions for creating weapons from a bedpost and a frying pan or a broom and a kitchen knife. Such images undercut the book's real-world utility and broaden its appeal beyond the gamified goals of the app.

In negotiating between real-world goals and a story-world narrative, Alderman draws on similar techniques to Max Brooks. Advice on weapons, getting 'fit for battle' and fortifying the home recall Brooks' *Survival Guide* and evoke similar tensions between individualistic success and collective concepts of social organisation. On the surface, *Zombies, Run!* offers a more positive endorsement of communal life with a whole section dedicated to 'building a community' and repeated advice to 'involve other people' (Alderman, 2016, 81–108). Some of this advice is directed vertically, since reporting your training routine successes to a buddy supports real-world fitness goals. Other aspects of Alderman's collective vision seem to be directed horizontally as she imagines a new world that might emerge from the rubble of a devastated capitalist society. In this world human life is valuable and communities should adopt the principles of 'truth and reconciliation', values that echo the libertarian socialist utopia of William Morris' *News from Nowhere* (1890). Frivolous consumer luxuries are gone along with our carefully curated 'social bubbles' (Alderman, 2016, 95). Instead, citizens are encouraged to develop equitable systems for sharing resources, play communal games to encourage bonding, and mediate conflict. However, this collective utopia is destabilised by the necessary individualistic ideologies inherent in gamification. As in Brooks' zombie world, success depends on self-discipline, a strong individual will and bodily capacity. You are advised to keep a regular schedule, set work tasks each day and remain 'useful and active' (Alderman, 2016, 41). Although the emphasis is on success for everyone, gamification evokes competition. In a story-world overrun by undead monsters, this takes on a sinister tone illustrated by 'Ms VP of Penrith': She realises that she cannot continue with her pre-apocalypse ways of laziness and poor eating: 'If there's anything that'll make you want to start exercising, it's seeing slower people get caught and eaten!' (Alderman, 2016, 65). This marks the limits of Alderman's communal vision and the utopian rhetoric of gamification for well-being more generally. The *Zombies, Run!* community is composed of able bodies and those with the means to live (and eat) well. Those without sufficient bodily capacity and self-discipline swiftly become one of the shambling, decomposing failures against whom the successful define themselves.

Alderman's gamified zombie world also confronts the problems of developing a participatory culture. Jenkins notes that collective efforts are needed to promote the kinds of literacy and capacities required if convergence culture is not to be hi-jacked by profit-seeking and commodifying corporate power. Collective efforts define participatory culture, which has low barriers to engagement; systems of informal mentorship – whereby what is known by the experienced is passed to novices; collaborative and playful methods of problem-solving and, finally, cultivates a sense of belonging and connection (Jenkins, 2006b). Alderman's collective society aims at developing the literacies and capacities Jenkins describes as germane to democratic participatory culture. For example, she shares important 'Zobo codes' found on buildings in the deserted cities. The codes warn fellow survivors of danger and provides information about the terrain. This new form of literacy allows the sharing of expertise with novices and develops social connection between disparate groups (Alderman, 2016, 60). Collective intelligence is also key, exemplified in the story of Maxine Myers, a former doctor who set out to find fellow survivors and 'teach other people what we knew' (Alderman, 2016, 78). These knowledge communities are not organised via a martial hierarchy as in Brooks; Alderman creates multiple case studies so there are multiple voices, all offering differing expertise and experience. The emphasis is on finding creative bottom-up solutions to problems of communication. For example, Alderman (2016, 106) tells readers how to access 'Rofflenet', a make-shift replacement for the internet described as a 'bulletin board from the 1970s' through which you can 'swap recipes, tips and techniques with other survivors'. Nonetheless, beyond the pages of the book Rofflenet is actually a pay-for members-only online service.

This discrepancy between the participatory culture imagined in the story world and the realities of for-profit gamification is a reminder of Jenkins' (2006a, 3) warning that 'not all participants are created equal'. Indeed, member access to the *Zombies, Run!* community is tiered: paying more also gets you extras including branded clothing and other merchandise. The horizontal address to a democratic and non-hierarchical community of survivors contrasts with the real-world inequalities inherent in gamification. In the book, Alderman provides an extensive appendix detailing how to build a 'Rofflenet Node' from scavenged tech, but, in reality, app users require 'iPhone 5s or above, running iOS 11.2+' or an 'android phone with at least 1GB RAM running Android 5+' ('Zombies, Run! Virtual Race FAQ', n.d.). Here, economic constraints undercut the utopian vision of the story world. Simultaneously, this discrepancy highlights the fact that claims about gamification as a transformative political tool elides the economic and technological barriers that Jenkins suggests hamper the development of participatory culture.

Conclusions

This analysis of ludified zombie fiction suggests that zombies are not simple metaphors for social anxiety nor political allegories, nor are they a blank target for violent gameplay. Ludification requires new and more complex readings of zombies. Moreover, zombies are not necessarily an easy fit with ludification and gamification. For one thing, the devastated world of a zombie apocalypse undercuts utility-oriented goals with not a little degree of nihilism. What is the point in setting lifestyle goals if you are going to end up a drooling monster? Likewise, gameful experiences identified by Koivisto (2017, 19) as germane to gamification, such as mastery and competence, are ironically undercut by the hero's gory demise. Ludified narrative structures suggest interactivity and participation, too, positioning the reader as a 'player'. However, this does not necessarily result in an increased sense of agency as the game space becomes a claustrophobic labyrinth with no obvious escape route. Indeed, zombie fictions bring the beleaguered concept of agency in games to the fore, disrupting reader and player-character identification to suggest that agency is not, after all, a subjective nor individual property.

In terms of political and social power, gameful interactions in the zombie apocalypse pit individualistic goals of competence and mastery against the promises of participatory culture and collective action. Likewise revealing tensions at the heart of convergence culture, narratives that imagine the decline of capitalist civilisation paradoxically work to promote commodification as players are encouraged to consumer merchandise related to the story world. Zombies are thus neither radical nor conservative but serve contradictory political ideologies. The political ambivalence of the zombie mirrors tensions in convergence culture, which functions neither to assert hegemonic power nor to liberate the economically oppressed. Indeed, in their dire consequences for failure, gamified zombie fictions reveal that gamification all-too often elides issues of gender, social class and (dis)ability in its promotion of self-improvement and well-being. In terms of the psychological explanations for gamification, the apocalypse seems to provide the immersion needed for intrinsic motivation, yet the presence of death undercuts utility-oriented goals, revealing them to be absurd. This discrepancy between the rhetoric of motivation and well-being clashes with images of despair, death and futility, allowing writers a degree of ironic distance on the process. Thus, zombies are disruptive figurations that reveal tensions in the broader cultural processes by which they have been co-opted.

Interactions and Sensoria

Fleischer, R. (Director). 2009. *Zombieland*, Sony Pictures Releasing (88 minutes).
Gilmour, J. and Vega, I. 2014. *Dead of Winter*, Plaid Hat Games.
Hubner, L., Leaning, M. and Manning, P. Eds 2015. *The Zombie Renaissance in Popular Culture*, Basingstoke, Palgrave Macmillan.
Livingston, I. 2012. *Blood of the Zombie* (*Fighting Fantasy*), London, Puffin.
PopCap Games. 2009. *Plants vs Zombies*, PopCap Games and Electronic Arts.
Romero, G.e.A. (Director). 1968. *Night of the Living Dead*, Continental Distributing (96 minutes).

References

Abbott, S. 2016. *Undead Apocalypse: Vampires and Zombies in the 21st Century*, Edinburgh, Edinburgh University Press.
Alderman, N. 2016. *Zombies, Run! Keeping Fit and Living Well in the Current Zombie Emergency*. London, Penguin.
Austin, E. 2015. Zombie culture: dissent, celebration and the carnivalesque in social spaces. In *The Zombie Renaissance in Popular Culture*, Eds L. Hubner, M. Leaning and P. Manning, pp. 174–191, Basingstoke, Palgrave Macmillan.

Backe, H.-J. and Aarseth, E. 2014. Ludic zombies: an examination of zombieism in games. In *Proceedings of the 2013 DiGRA International Conference: DeFragging Game Studies*. Retrieved from http://www.digra.org/wp-content/uploads/digital-library/paper_405.pdf

Bishop, K.W. 2015. I always wanted to see how the other half lives: The contemporary zombie as seductive proselyte. In *The Zombie Renaissance in Popular Culture*, Eds L. Hubner, M. Leaning and P. Manning, pp. 26–40, Basingstoke, Palgrave Macmillan.

Bogost, I. 2011a. Persuasive games: exploitationware. *Gamasutra: The Art and Business of Making Games*, 3 May. Available at: https://ubm.io/2pW2DcQ

Bogost, I. 2011b. Gamification is bullshit, *The Atlantic*, 9 August. Available at: http://bit.ly/2WkHZiF

Brailler, M. 2011. *Can You Survive the Zombie Apocalypse?* New York, Gallery Books.

Brooks, M. 2004. *The Zombie Survival Guide: Complete Protection from the Living Dead*, London, Gerald Duckworth and Co.

Buckles, M.A. 1985. *Interactive Fiction: The Computer Storygame 'Adventure'*. PhD thesis, San Diego, CA, University of California.

Deterding, S., Dixon, D., Khaled, R. and Nacke, L. 2011. From game design elements to gamefulness: Defining gamification. In *Proceedings of the 15th International Academic MindTrek Conference: Envisioning Future Media Environments*, Tampere, Finland. 9-15. Available at: https://doi.org/10.1145/2181037.2181040

Fuchs, M., Fizek, S., Ruffino, P. and Schrape, N. 2014. Introduction. In *Rethinking Gamification*, Eds M.S.F. Fuchs, P. Ruffino and N. Schrap, pp. 7–20, Lüneberg, Meson Press.

Koivisto, J. 2017. *Gamification: A Study on Users, Benefits and Literature*, Tampere, Tampere University Press.

Krzywinska, T. 2008. Zombies in gamespace: form, context, and meaning in zombie-based videogames. In *Zombie Culture: Autopsies of the Living Dead*, Eds S. McIntosh and M. Leverette, pp. 153–168, Lanham, MD, Scarecrow Press.

Jenkins, H. 2006a. *Convergence Culture: Where Old and New Media Collide*, New York, NY, New York University Press.

Jenkins, H. 2006b. Confronting the challenges of participatory culture: media education for the 21st century (part one). *Confessions of an Aca-Fan*, 19 October. Available at: http://bit.ly/2PqjCyw

Jenkins, H. 2019. Participatory politics in an age of crisis: Henry Jenkins & Nico Carpentier (Part II). *Confessions of an Aca-Fan*, 3 June. Available at: http://bit.ly/2qRew4j

Lauro, S.J. 2011. Playing dead: zombies invade performance art ... and your neighbourhood. In *Better Off Dead: The Evolution of the Zombie as Post-human*, Eds D. Christie and S.J. Lauro, pp. 205–230, New York, NY, Fordham University Press.

MacCallum-Stewart, E. and Parsler, J. 2007. Illusory agency in vampire: the masquerade – bloodlines. *Dichtung Digital*. Available at: http://bit.ly/32TZiK2

Mäyrä, F. 2008. *An Introduction to Game Studies: Games in Culture*, London, Sage.

McGonigal, J. 2011. *Reality is Broken: Why Games Make Us Better and How They Can Change the World*, New York, NY, Penguin Press.

Murray, J. 1997. *Hamlet on the Holodeck*, New York, NY, Free Press.

Platts, T. 2013. Locating zombies in the sociology of popular culture, *Sociology Compass*, 7(7), 47–560. https://doi-org.ezproxy.mmu.ac.uk/10.1111/soc4.12053

Raessens, J. 2014. The ludification of culture. In *Rethinking Gamification*, Eds M. Fuchs, S. Fizek, P. Ruffino and N. Schrape, pp. 91–118, Lüneberg, Meson Press.

Raessens, J. 2006. Playful identities, or the ludification of culture, *Games and Culture*, 1(1), 52–57. https://doi.org/10.1177/1555412005281779

Six to Start and Naomi Alderman. 2012. *Zombies, Run!*, Six to Start.

Tenga, A. and Zimmerman, E. 2013. Vampire gentlemen and zombie beasts: a rendering of true monstrosity, *Gothic Studies*, 15(1), 76–87. https://doi.org/10.7227/GS.15.1.8

Wake, P. 2016. Life and death in the second person: identification, empathy and antipathy in the adventure gamebook, *Narrative*, 24(2), 190–210. http://doi.org/10.1353/nar.2016.0009

Wardrip-Fruin, N., Mateas, M., Dow, S. and Sali, S. 2009. Agency reconsidered. In *Proceedings of the 2009 DiGRA International Conference: Breaking New Ground: Innovation in Games, Play, Practice and Theory*. Retrieved from http://www.digra.org/digital-library/publications/agency-reconsidered/

'Zombies, Run! Virtual Race FAQ.' n.d. Available at: https://zombiesvirtualrace.com/faq [Accessed on 23 June 2019].

Part 4

Performing Playful Realities

Chapter 11

The Jovial Aesthetics of the Death-Positivity Movement: Notes on the Appeal of Playfulness in Activism

Solveiga Zibaite

Abstract

The death-positive movement can be described as a de-centralised contemporary social movement originating and operating predominantly in the global West, specifically the United States, connecting death workers, educators, artists, journalists, etc., and geared towards encouraging open dialogue about death and dying. It has succeeded in capturing significant media attention over the last few years and is largely driven by its strong social media presence. This chapter looks at 'playfulness' within the death-positive movement. Examining the dimension of 'playfulness' addresses the affective aspect of communication that in this movement is inseparable from the message. First, the author investigates the aesthetics of representation through death-positive merchandise, produced and advertised by *The Order of the Good Death's* (subsequently – The Order) core members. Second, the author considers some of the cultural output produced under the umbrella of death-positivity, but not by the core movement members, specifically taking the first video game to be explicitly marketed as death-positive – *A Mortician's Tale* (Laundry Bear Games, 2017) as a case study. Finally, the author analyses the role of entertainment value in the movement's leaders' discourse on death, taking leader of The Order Caitlin Doughty's playful rhetoric on her YouTube channel, Twitter profile, and Instagram pages. The manifesto, found on the movement's official website (http://www.orderofthegooddeath.com/) encourages its participants to break the 'culture of silence' around death, indicating that the whole premise of the movement is based on the supposed presence of death denial in Western countries. Ultimately, the author argues that by eliciting playfulness, this

Death, Culture & Leisure: Playing Dead, 157–172

doi:10.1108/978-1-83909-037-020201018

challenge to the social climate becomes a somewhat jovial and enjoyable endeavour and generates response from outside the movement.

Keywords: Death-positivity movement; humour; activism; entertainment; dialogism; aesthetics

Introduction

This chapter unpacks the role of 'playfulness' in the creation, sustaining, and promoting the death-positive movement's mission. By using the term jovial aesthetics, I refer to representations of the movement in images and speech that are available for consumption. Namely, the humorous rhetoric present on social media and in educational content, and death-positive merchandise, which are the playful ways this movement engages with itself and its consumers to facilitate understanding of the movement at its core. Examining 'playfulness' addresses the affective aspect of communication that is inseparable from the message. This analysis illuminates the workings through which the death-positive movement becomes inclusive for some and alienating for others. In this chapter I consider three aspects of analysis. First, I investigate the aesthetics of representation through death-positive merchandise, produced and advertised by *The Order of the Good Death's* (subsequently – The Order) core members. Second, I consider a selection of the cultural output produced under the umbrella of death-positivity, but not by the core movement members, specifically taking the first video game to be explicitly marketed as death-positive – *A Mortician's Tale* (Laundry Bear Games, 2017) as a case study. Lastly, I also analyse the role of entertainment value in the movement's leaders' discourse on death, taking the movement's spearhead's Caitlin Doughty's playful rhetoric on her YouTube channel, Twitter profile, and Instagram pages. It is important to remark that while this chapter focusses on the humorous rhetoric and images utilised to promote the movement and engage with followers, the death-positive movement simultaneously exhibits a poignant and well-articulated goal-oriented mission.

The death-positive movement can be described as a de-centralised contemporary social movement originating in the United States, connecting death workers, educators, artists, journalists, etc., and geared towards encouraging open dialogue about death, dying, and bereavement. It is mainly concerned with changing the relationships of what is broadly understood as Western culture, with death and dying. The definition of the Western world includes mainly the United States and a broad sweep of cultures where commercial, professionalised funeral industry has come to dominate how death is handled. The Order is the self-fashioned core for the contemporary death-positive movement, founded in 2011, with the term 'death-positive' coined in 2013 by its originator and spearhead, millennial LA-based mortician Caitlin Doughty. One of Doughty's simultaneously most tangible and most elusive aims is to reform Western funeral industry practices. Doughty owned a non-profit funeral home *Undertaking L.A.* from 2015 until 2019, promoted family-led care of the dead, as well as endorsed natural burials, and minimising the impact of burying practices and products on nature

(Undertaking LA: http://www.orderofthegooddeath.com/undertaking-l-a). She wishes to eradicate the fear of human corpses being dangerous and polluting for the living. The other core members include official content producers Louise Hung, Sarah Chavez, Director of The Order, Megan Rosenbloom, Director of Death Salon, and Amber Carvaly, Director of Undertaking LA. The death-positive movement has lobbied for laws such as the legalisation of aquamation (a process for the disposal of human and pet remains using lye and heat), emerging greener death technologies, such as body composting, and an End of Life Option Act in the US. It has lobbied against bills, such as prohibiting families to keep their dead loved ones at home for more than 48 hours. However, these clearly delineated initiatives are not immediately visible if one's only glance at the movement is the way its image is projected and managed on social media. These platforms are a significant vehicle for the promotion of the movement and have high interaction levels with adherents all over the world. The death positive movement has also captured significant media attention over the last few years, with features in wide-ranging outlets, from *Dazed* (Gilmour, 2016), *Marie Claire* (Ortiz, 2016) to *The Guardian* (Kelly, 2017) and the *New York Times* (Leland, 2018). These publications all aim to highlight to the general readership that death-positivity does not entail a morbid fascination with death, but encourages an honest acknowledgment of the reality of death. Another common aspect is that all of these, and many not mentioned publications, enthusiastically note that the movement is populated and propagated mostly by women.

The movement's core members continuously give exceptional attention to recognising the complexity of emotions that are associated with the phenomena of death and dying:

> There is no hierarchy of emotions here. A person can feel in full control of their fears one day and lose a friend or family member the next. I love using humour to engage with death. That doesn't mean I won't be crying the next morning. Death contains multitudes, just as humans contain multitudes, just as emotions contain multitudes. Life is a wild journey, and death positive means riding the never-ending waves, not climbing a single peak. (Doughty, 2018)

However, as is also evident from the quote above, humour is a significant part of the way the movement chooses to present itself.

Death and Dark Humour

Gallows humour is a well-known sociological phenomenon that has been explored extensively (Obrdlik, 1942; Palmer, 1994; Pasquali, 2003 among others). Obrdlik (1942) conceptualises it as 'type of humour that arises in a type of precarious or dangerous situation' (Obrdlik, 1942, 709). Utilising their experience of gallows humour in Czechoslovakia during Nazi occupation, they found it to have significant positive effect on the morale and spirit of resistance of people existing in absolute uncertainty about their livelihoods. Obrdlik (1942, 716) locates the specificity of gallows humour in its intentionality – its 'humour with a definite purpose – that is, to ridicule with irony, invectives, and sarcasm in order to become a means of an effective social control'. By means of ridicule, gallows humour also undermines the authority of the oppressor (Billig, 2005). Gallows humour has also been explored in relation to death and dying (Wear et al., 2006; Watson, 2011; Cain, 2012). Thorson (1985, 206) reviewed studies in humour and death and found that

> a number of scholars have seen the humour associated with death as a high-level defence mechanism, a means to deflect the pain of death or grief, a means of social control, and a means of coping with grim realities.

Thorson proposes that humour can actually be an 'offense' mechanism against death and that by laughing at one's own finiteness, a sense of control is gained. He writes: 'while it cannot deny the reality of death, it can deny the importance of the reality of death' (Thorson, 1985, 206). Knickerbocker (1964) defined gallows humour as 'terrible and marvellous at the same time'. Gallows humour as a tool for individual defiance, social mobilisation, temporarily suspension authority or reality, promoting camaraderie through outrageousness and ridicule can be explored further as a kindred notion to Bakhtinian carnival. Mikhail Bakhtin (1875–1975) was a Russian literary theorist and one of the key discourse theorists of the twentieth century. Sullivan (2012, 13) writes that due to its remarkable fluidity, Bahktinian dialogism in qualitative research has been appropriated by discourse analysts, narrative researchers and in interpretive phenomenological analysis. Arguably the most widely used Bakhtinian concepts are 'the carnivalesque' and 'heteroglossia'. Carnivalesque refers to an event or behaviour, characterised by distortion and disorder, wherein social class and status are reversed, laughter is corporeal and communal, and the grotesque is celebrated. Heteroglossia, introduced in *The Discourse of the Novel* (Bakhtin, 1981) is the multi-layered nature of language, which is inhabited not only by social dialects, technical, professional jargons, fashions, but also socio-ideological contradictions inherited from the past and carried forward. As the death positive movement dangles on the ledge of order and charismatic chaos, a tangible mission for 'change' and inverted meanings, serious matters and sarcastic satire, it is particularly suitable to be examined through this paradigm. Hoy (1992, 766) sees it essential to apply Bakhtinian analysis to contemporary magazines, comedy, advertisements, popular music, art, and fashion,

> since in their continual interchange and deliberate fusion of high and low styles, politics, parody and pastiche, comic strip and literature, haute couture and street fashion, they constitute a singular shifting dialogism whose rich carnival of discourse lies open to Bakhtin's radical definition of 'novelness'.

Hoy accentuates intentionality of mixing language, destabilising the usual, similarly to how gallows humour operates.

Within the remainder of this chapter, I demonstrate that while the death-positive movement is not a mainstream movement, it is continuously engaging with the tropes of popular culture, parodying itself and the critical spectator. Kutz-Flamenbaum (2014, 295) remarks that compared to a wide interdisciplinary body of work on humour in general and the assertion that humour is historically widespread within both social movements and political culture, the literature on humour within social movement scholarship is thin. Among most important contributions he mentions Speier (1998), Tuene (2007), Swart (2009), Boyer and Yurchak (2010). Explicitly worth mentioning are works on protest (Bruner, 2005, Hart, 2007, 2016, Davies, 2007) and political parody (Kowsar, 2012; Petrović, 2018). This chapter expands the research field by providing an analysis of humour as strategically used in a largely internet-based and interactive social movement. Göttke (2015, 30) argues that the discussion of the entanglement of humour and violence in political protest, where effigies are burned is enabled precisely because Bakhtin's concept of the carnivalesque uniquely combines an aesthetic (the grotesque) with a strategy (reversal and debasement) and the effects of this practice on the social body. The death-positive movement is dedicated to changing Western funerary practices and the relationship with death and the dead body, thus I claim that it is a form of long term protest, a carnivalesque endeavour wherein aesthetics are deliberately confrontational. Via a momentary suspension and replacing of what the movement's proponents see as the conventional sombre and muted emotional attunement to matters of death and decay, with 'an overwhelming – even sacred – sense of collective camaraderie' (Turner, 1974, 20–59). These themes are interwoven throughout this chapter but ultimately help advance the contention that the use of humour in the death-positive movement contributes to the collective identity and solidarity. Humour aids the construction of the death-positive movement as accessible, tongue-in-cheek, playful, and an ambiguous endeavour.

However, questions concerning audience inclusivity become paramount. The death-positive movement is spearheaded predominantly by young white females. It is, however, very much concerned with the livelihoods of marginalised communities, as evident from some of their blog posts, such as *Dying Queer in 2018* (8 August 2018), *Considerations for Caregivers in marginalised communities* (20 December 2018) Playfulness and humour in the death-positive movement in its conventional media apparitions act as a resource for social mobilisation, which is especially attractive to younger demographics. Alejandra Trevino, a senior at North High School in Ohio, is described in her biography at TEDxAkron in 2018 as 'an advocate for death positivity for all people and cultures' (Cottom, 2018). She argued that death positivity 'for the most part, is not really about your death. It's more about making your life mean something' (Cottom, 2018). This rhetoric is often the case in talks from young death positivity advocates. Their claims seem to act as a justification for an interest in death that have the potential to reduce the death positivity movement reduced to a memento mori with a sparkly pop-culture twist, carrying a YOLO-esque (you only live once) essence. I argue that a lot of the work of the death-positive movement is about changing attitudes towards death and dying from a young age and playfulness in this case makes maximising inclusion of different demographics problematic. Furthermore, there is an element of imagined community and surface participation: even though it is claimed that all it requires to become a member of The Order are a few clicks of a mouse to put down one's name in a long list of other *deathlings*, as Doughty (2018) endearingly calls her followers. The Order aims to showcase the people at the forefront of the change

> from a designer who creates clothing and shrouds that decompose at the same rate as your corpse, to the funeral director opening a funeral home that combines an art and culture space with a place for the dead. (The Order of The Good Death: About)

Deathlings thus are aligned with the movement when they *consume* its message – assimilating the humorous rhetoric, or literally, consuming the product – merchandise, which I move on to discuss next.

Fashion and Aesthetics

> FUTURE
> CORPSE
>
> FUTURE
> CORPSE
>
> FUTURE
> CORPSE

These words form a slogan on a t-shirt sold on The Order's Etsy page (https://www.etsy.com/shop/TheGoodDeath). The store's web front promotes 'death-positive goodies & more'. A list of item categories appears for you to choose from: 'clothe your corpse', 'accessorise your bones', 'grave décor', and 'flesh care'. On The Order's Facebook page, I came across a picture uploaded in the comments of a *deathling* showing off her 'future corpse' t-shirt in a selfie, proudly exclaiming how terrified of her a gaggle of silver-haired ladies at Walmart were. Caitlin Doughty's Instagram page (@thegooddeath), which acts as the official Order's Instagram, is abundant with images of Caitlin and her daily activities, promoting death-positive merchandise, events, and fan art usually accompanied by elaborate humorous captions like: 'disembodied embalmed hearts are the cherry on top of the death cake'[1] or 'Oops I did it again. Picnicking with a Tule Elk skull. What's your favorite death-related activity you did this weekend?'[2]

The examples of engagement between The Order's official channels and *deathlings* in this paragraph exhibit a certain rhetoric that might appear grotesque for an unsuspecting internet passer-by coming across any of these social media platforms by accident. It is necessary here to examine the role of death-positive merchandise in both creating and maintaining a death-positive identity as a statement

[1]@thegooddeath, Instagram, 26 August 2018.
[2]@thegooddeath, Instagram, 1 July 2019.

through personal style and consumption choices. Death-positive merchandise is chosen for analysis, because it is consciously produced by stakeholders and available for consumption as an outward sign of membership and support for the movement. Both producers and consumers co-construct the meanings of the movement, the former by selecting messages to broadcast as representations of their mission and the latter as finding the message worthwhile to be spread by broadcasting it through their individual appearance. And this arena is precisely where conflation of activism and entertainment happens.

A classic black t-shirt with a simple light grey slogan 'my corpse my choice' is displayed alongside a white vest with a Grim Reaper on a pink surfboard, grinning in the sunset background. The blue wave underneath the reaper spells 'death positive'. This discrepancy in tone is disorienting. It is even more disorienting when considered against the 'corpse meditation' t-shirt, which poignantly illustrates the stages of decomposition, or the newest addition to the merchandise at the time of writing – a t-shirt with the imagery of saint Santa Muerte, who is an 'unofficial patron saint of LGBT persons' with the caption of *Loved for exactly who I am.*[3] Are these to be worn on different occasions? Should we think about death constantly, thus have death-positive t-shirts for all times of day? The movement appears to be a shapeshifter – a social justice movement by day and an Instagram hashtag opportunity by night. The shock factor of the movement's iconography might have to do with the proposed need to come face to face with the gruesome reality of death, and that can be seen successfully applied in 'corpse meditation' t-shirt, but the dissonance between the messages and imagery in available t-shirt designs heightens this promotional endeavour to a level of almost surreal allegory. This uncertainty obstructs broadcasting a clear message of the movement for the part of proponents who are both consumers and representatives. This is also coupled with Doughty's penetrating charisma which is widely utilised. Stickers with cartoon versions of the core members – all female - are available for purchase. They possess infant looking faces and lustrous manes that seems to be a very prominent motive in their branding. Doughty's own late cat The Meow appears on new 2018 t-shirt designs, a minimalist 'cute' rendition of it on its back, playing with a skull in its paws. Inclusion of a symbol from Doughty's personal lore further strengthens the reins that Doughty holds on this movement.

References to fleshiness and human corporeality and inevitable decay in product names in TheGoodDeath Etsy store are numerous. Consider *Lip Em'balms* with such names as *Decomp'uccino* and *Butter off Dead*, or a selection of body butters marketed as *Adipocere,* otherwise known as corpse wax, an organic substance formed post-mortem by the anaerobic bacterial hydrolysis of fat in tissue. This merchandise was being produced by Doughty's own funeral home Undertaking L.A., the closing of which also resulted in a suspension of the Etsy store for the time being. The references to corporality correlate with the focus on promoting agency in making post-mortem decisions, accustoming ourselves with the processes of decay and the desire to normalise such topics and conversations. On

[3]@thegooddeathstore Instagram, 15 June 2019.

Patreon, the crowdfunding website that allows 'patrons', who pay a monthly fee to support creators in exchange for early access or additional content, Doughty offers an 'Iconic Corpse tote bag' and in April 2019, for a limited time, it was possible to receive an original, Patreon-only sticker of Jeremy Bentham's preserved head. This otherwise unavailable merchandise affirms the feeling that the supporters are co-creating the movement and are able to have evidence of this exclusive membership and support. In her work on images of morbidity in the fashion world (skull, skeleton imagery, 'corpse chic' seen in fashion editorials), Foltyn (2010) notes that 'in the contemporary world of style, death is worn as a prop in the performance of fashion'. While there she refers to the wider global fashion trends, the following stark warning she issues is relevant to here discussed niche consumption of death positive merchandise:

> as we naively wear or gaze upon these often beautiful spectacles of death, we may forget that one day death will come for us too, and its arrival won't be a style statement. (Foltyn, 2010)

Instagram search for #deathpositive returns over 25,000 results, showcasing various fashion and nail art choices (notable example of the hashtag from TheGoodDeath's Instagram page is of nail art done with real mouse and vole bones) as well as Victorian memorabilia, *memento mori*, stylised and anatomical skull images. The humorous, somewhat shocking and borderline grotesque messages of death-positive merchandise also provide opportunity to build a sense of collective identity through shared distaste of a perceived culture of denial of death and to showcase membership through a shared vocabulary. Clothing, accessorising in death positive merchandise is broadcasting the message using the most personal canvas. A part of the sustainability of the death-positive movement lies in its focus on the empowerment of the individual. This emphasises the power of the individual to subverse, to be ironic, and confrontational from the first glance, to not play along the rules. However, another question quickly arises in regard to personal expression of death-positivity – *is* there a way a death-positive person *should* look or generally looks? In January 2018, Doughty published a dedicated article on the Order's website, delineating 'what the death-positive movement is not'. One myth that Doughty seeks to dispel is connected to the appearance of the Order members, often styling themselves in darker, alternative fashion. A critical blog post 'The Politics of death and dying: A challenge to the Death Positive Movement' in the Death Rookie blog, posted on 26 September 2018, states that 'death acceptance conversations (among non-clinicians) seem dominated by young white women with a kicky interest in Victorian death tchotchkes'. A blog about the movement, simply called The Death-Positive Movement blog, in a post 'What's Missing?' (Death-positive movement blog, 2016) critically states that by embracing a morbid persona and stereotypical style, Doughty 'queers the very thing she want to normalize'. Interestingly, when attempting to dispel the myth that the way death positive people are stereotypically perceived to dress somehow affects their integrity, Doughty uses an identically dismissive contra-stereotype in a blog post What Death Positive Is NOT' posted on 25 January 2018: 'Must everyone wear tweed and elbow patches to prove their bonafides as activists?'. In the following sentences, she goes on to quote her fellow Order member Sarah Chavez, who says, that this line of thinking is a convenient way to "dismiss and diminish the voices of women in the movement by labelling them 'goth'".

The furthering of this stereotype can be seen in *A Mortician's Tale* (Laundry Bear Games, 2017), the first video game, created and explicitly marketed as death-positive. It was released on 18 October 2017 and received mixed reactions from reviewers and players. It is played through the perspective of a recent mortuary science graduate Charlie who joins a family run funeral business. In its short gameplay (around one hour), she deals with eight bodies with different backstories. You read an email, you prepare the presented body, attend a wake, repeat- all with a click of a mouse. Charlie is an attractive tattooed raven-haired female, in the presentational poster seemingly blushing as she gazes upon a pair of large scissors. It bears a passing resemblance to Doughty, but then it should be explicit that it refers to a specific person, because otherwise, in this aspect, the game appears to perpetuate the pop-culture stereotype of females, working with death (e.g. in the funeral industry and in crime) as quirky, seductive and virtuously not aware of it, and, most importantly, 'alternative' (often gothic, tattooed), adding to the explicitly feminine and fetishised renditions of death and death work. While beauty and death work are not mutually exclusive in any way, it becomes problematic when an extremely aestheticised image is consciously selected to represent death work in an already small pool of representations.

Interestingly, the artwork for the movement's leaders' new podcast *Death in the Afternoon* makes a point to showcase the cartoon skulls of Caitlin Doughty, Sarah Chavez, and Louise Hung that are recognisable because of their nearly identical glossy hairstyles besides slightly different fringe shapes. This is also a very ambivalent decision. While it can be seen simply as marketing a familiar face to the existing fanbase and helping make connections for the newcomers – a strategy for expansion, it could also be perceived as a tongue-in-cheek play and exaggeration of a trope that the majority of death positive activists have long hair and bangs, which the movement spearheads also playfully engage in. For example, in Red Ice podcast titled 'Dinesh D. Souza is Wrong and the Death Positive Movement', aired 4 September 2017, the hosts remarked that the leaders of the death-positive movement, Doughty and her colleagues, which are all young and female, 'look like bad Betty Page caricatures and that the way they look and dress is a way of accurately pathologising someone'. Unapologetic exaggeration of the similar features of the core members' appearances in marketing materials can be considered a response to such and similar unwarranted and prejudiced criticism. The leaders can draw on such criticism and co-respond to it: 'The Order of the Good Death is not a cult! Just because we have these glasses and bangs and sweaters and blank expressions doesn't mean anything. Stop saying that!'[4] The Twitter thread continues with Sarah Chavez posting a picture of a PowerPoint slide presentation: 'What's up with the bangs? Is it part of the initiation ritual?'[5]

---

[4]@TheGoodDeath, Twitter, 26 March 2018.

[5]@sarah_calavera, Twitter, 26 March 2018.

A dialogic encounter is present in all discussed instances of ambivalence of the humorous rhetoric and representation aesthetics. When Doughty and her colleagues play up the trope of them all having the same haircut, they are engaging in 'double voiced discourse' (Bakhtin, 1984, 94). This term contends that when a person speaks, they have an anticipated intention of being heard by a different discourse, already being aware of the critical concern or agenda, thus altering their language in a way to address the anticipated conversationalist (Bakhtin, 1984). As Baxter writes (2014, 3)

> this responsiveness goes well beyond normal conversational interactivity, and usually contains a 'power' dimension in that double-voicing can be used as a strategy to resist threats from more powerful others, to silence someone else, or even, to give someone a voice.

The statement 'we are not a cult', appearing as a tweet without a specific prompt is already a response to an imaginary interlocutor (not a specific person, but a version of a dominant discourse), oriented towards the representatives of dominant discourse or critics, as well as towards the members of the death-positive movement in order to affirm their relationality. In the rhetoric of the death-positive movement, the sideways glance is partly directed at itself. 'This kind of language is self-referential, self-regarding – aware, in a way, of its own shifting dialogism' (Hoy, 1992, 770). If the death-positive proponents are being constantly dismissed as morbid gothy millennials, they are repurposing this direct discourse by using it first on themselves to mock all involved and make its primary purpose, which is to discredit their legitimacy, lose its power. The aesthetics of the death-positive movement, discussed here in terms of merchandise available for consumption and an artistic social media content, thus consist mostly of a self-aware and self-ironic glance towards itself, with the exception of *A Mortician's Tale*. While it is not a creation of the core members, the game's designer Gabby daRienzo claims to be very closely aligned with the movement and has been inspired by Caitlin Doughty's book *Smoke Gets in Your Eyes*. It is explicitly marketed as part of the movement and is easily searchable online using these terms, and thus I argue that it merits being included in core cultural death-positive output so far – it represents the movement for anyone who looks for information about it online. I argue that the game's imagery and story does not possess this kind of reflexivity and self-irony and is the direct application of the surface death positive tropes – 'gothy', feminine death work.

Having considered imagery, I will now turn to reflect on the self-aware and self-ironic language that manifests in Doughty's rhetoric online, in terms of what Bakhtin conceptualised as the language of the merry rogue, showing a multi-layered perception of oneself and the culture climate the movement is operating it.

The Language of the Merry Rogue

The merry rogue in Bakhtin's world, coming from the archetype of the fool, uses his playful, self-conscious language to contrast the seemingly taken-for-granted language of the king to undermine the king's authority and disseminate opposing discourses to the hegemonic king's language. A lot of the death-positive movement's humour is driven by Caitlin Doughty's charismatic leadership. It is important to consider Doughty's own contemplations of use of humour to further her goal to reform Western death practices. Enlivening and inclusive education is a key part of her 'Ask a Mortician' videos on the online platform YouTube and humour is employed by Doughty for didactic purposes. Her videos, addressing a range of death-related issues, some with historical dimensions to them have gained widespread popularity and as of October 2019 has gained over 94 million views. On 11 February 2019 Doughty did not miss the opportunity to share on Twitter (@TheGoodDeath) that her YouTube channel has reached 666K subscribers, with a witty comment: 'Waiting patiently for this auspicious moment. Blessings upon the channel'. It seems that here Doughty plays with the prejudice that if people possess an interest in deathly matters, it is because they are interested in the occult. This continuous playful engagement with symbols related to morbidity in popular culture is another way she chooses to market her brand, working in the framework of awareness that the movement is susceptible to being perceived in popular culture as a macabre and morbid millennial subculture. In a short interview 'How Caitlin Doughty is Rethinking Death', when asked what she would like to change in the way we deal with death, Doughty replies:

> A lot of the way I talk is humorous and, hopefully, accessible. And that's because death gets a very bad rep, and a very sinister rep and by making it more humorous and open, and tragedy and comedy going together we can say, ok, its safe to have this conversation, it's a safe space, let's all just get in the pile and talk about it. (Santa-Maria, 2019)

In social media posts, especially in captions accompanying pictures of a new batch of merchandise, or under a shared piece of news, Doughty's lashings of humour are generous.

On October 9 Doughty posted a picture of Portland Mercury headline 'Death-Positive Mortician Caitlin Doughty Wants You to Know the Grim Reaper's Hand Is Up Your Butt' on her Instagram page and captioned it: 'this is a real headline in today's @Portlandmercury. To be fair, it's not clickbait, I do want you to know this'. The fact that she deems the confirmation that the headline is indeed 'real' post-worthy already signals a reflective awareness of the crassness and the shock value the headline exhibits. Then she makes an attempt to deny the sensationalism, but when she writes that she 'wants you to know this', it comes back to affirming the original impression that the headline makes. She could have just posted a picture of the headline with the same effect, but such engagement with the headline in the caption demonstrates a high degree of awareness about the way she herself chooses to drive the movement, that is, in an intrinsically humorous and ambivalent way. This is accompanied by unexpected and inconsistent shifts in rhetoric on the Order's social media. On 5 Februarys Instagram post, captioned 'Beauty Corner with Caitlin (unicorn emoji, whale emoji, applying nail polish emoji)', Caitlin is seen wearing a sheet mask with a face of a narwhal on it.

> Hi everyone, welcome to beauty corner with Caitlin. Tonight we are going to be reviewing the narwhal face mask with brightening vitamin C. My review is as follows. Did you know that the word narwhal, comes from the old Norse 'nar', meaning corpse? The reason the narwhal was thought to be a corpse is that they bobbed silently on top of the water not moving and had a pigmentation that was mottled, much like a drowned sailor. Thanks for watching! (waves). (@thegooddeath Instagram page, 6 February 2019)

She parodies beauty gurus by using an exaggerated monotone voice tone, expressionless face, bringing the package close to the camera, while covering her face so that the camera focusses on it. In this video she successfully blurs distinctions between social media influencer and activist, the movement's mission of education and entertainment value – in this instance it is neither one of those. Or in fact, both, as the phenomena of far-reaching internet celebrities has been proliferating and also well researched (Senft, 2008; Marwick, 2013; Abidin, 2018 among others). Cassegard (2012, 1) writes that 'play is not the opposite of reality, as is often asserted, but the opposite of powerlessness'. The impersonation of 'high priestess' of death, an *influencer* of sorts, exhibits a sideways glance that is however, directed at oneself. Andrew Boyd (2002) argues that irony has become a tool for new generation of activists. At first glance impenetrable and macabre statements and references might very easily confuse people who accidentally stumble across the movement's social media platforms and there is an undeniable danger of losing those potential *deathlings* who are not willing to play along with the humorous rules and engage in unpacking all the sideways glances. Appealing to mainstream audience seems to be unachievable in that way, however, Branagan (2007, 479) argues that 'humour involves a transmutation from negative hectoring to positive communication which others can enjoy, in an enriching process for both audience and actors'.

In the death-positive manifesto, found on The Order's website (http://www.orderofthegooddeath.com/death-positive), the participants are encouraged to break the 'culture of silence' around death, indicating that the whole premise of the movement is based on the supposed presence of death denial in Western countries. This is very clearly visible in one of the statements, taken from the movement's mission: 'death itself is natural, but the death anxiety and terror of modern culture are not'. The Order is against using euphemisms to describe death and dying, such as 'passed away', 'went home', etc. They show awareness of the language that can be associated with Bakhtin's notion that laughter has the capacity to strip the object of the false verbal and ideological husk that encloses it. The language key movement figures use in social media and educational content when talking about the practicalities of human death and decomposition is blunt and straightforward, while language about the movement itself is humorously self-referential. This combination of expression can be conceptualised as a non-threatening, non-violent challenge to the hegemony of the vision of death denying culture, due to its fluid nature and the visibility of core leaders easily susceptible to major criticism, but persisting nevertheless as a critique of the lack of death awareness in contemporary North American culture.

Conclusion

The death-positive movement utilises a wide range of avenues to promote their educational mission – workshops at their L.A. funeral home, the biannual Death Salons, Ask a Mortician videos, abundant social media engagement. With this comes the variety of uses of language and discourse. This chapter has focussed on the playful ways the movement engages with itself and its consumers – humorous rhetoric and the use of symbols in various representations of the movement that facilitate understanding of the movement or are available for consumption. I deemed using Bakhtinian concepts of the carnivalesque and heteroglossia especially suitable to interrogate the issue because of the movement's primary nature of a protest against existing Western funerary practices and relationship with death and decay while seemingly playing along with their possible perception in wider society as morbid, cliquey and not serious.

The death-positive movement provides both neutrally articulated educational content on their website and in their political work (lobbying for and against laws), and humorous chaos, led by charismatic leader Caitlin Doughty, who I referred to as the merry rogue. The rhetoric I examined on their social media platforms is a fusion of colloquial language, didactic, and educational rhetoric, technical jargon and puns. The marginal death-positive has harnessed the power of social media and is thriving on this kind of engagement with its followers. However, due to the inconsistencies in the tone of the content, the stance that the movement takes might appear shaky for an outside observer. For future work, it would be fruitful to discuss how this humorous presentation of matters of death and decay makes us reconsider Lofland's (1978) influential suggestion that death and dying can become a fad or a fashion. This statement has served as a critique of the death awareness movement in the 1960s and 1970s, referred to by Lofland as 'Happy Death Movement', supposedly offering a positivist view of death that romanticises dying and overemphasises emotional expressiveness as therapy for the dying and bereaved. The death-positive movement, I claim, has harnessed the power of humour to make people 'warm up' to things – mortality in this matter. It might be a fashion, but it is a critical fashion then. By eliciting ambivalence between the mission of the movement and its casual humorous representations through playfulness, this movement, a challenge to the social climate, becomes a somewhat jovial and enjoyable endeavour and is able to generate an effective response from outside the movement, while being aware of the potential misunderstandings and shortcomings of this playful approach to communication.

Interactions and Sensoria

Death in the Afternoon. n.d. *Death in the Afternoon*. https://deathintheafternoon.libsyn.com/ [Accessed on 30 April 2020].

Doughty, C. n.d. *Ask A Mortician*, YouTube. https://www.youtube.com/channel/UCi5iiEyLwSLvlqnMi02u5gQ.[Accessed 30 April 2020].

Laundry Bear Games. 2017. *A Mortician's Tale*. Laundry Bear Games.

The Order of the Good Death. n.d. *The Order of the Good Death*. http://www.orderofthegooddeath.com/ [Accessed 30 April 2020].

References

Abidin, C. 2018. *Internet Celebrity: Understanding Fame Online*, Bingley, Emerald.

Bakhtin, M.M. 1984. *Rabelais and His World*, Translated by H. Iswolsky, Bloomington, IN, Indiana University Press.

Bakhtin, M.M. 1981. Discourse in the novel. In *The Dialogic Imagination*, Ed. M. Holquist, pp. 259–442, Translated by C. Emerson and M. Holquist, Austin, TX, University of Texas.

Baxter, J. 2014. *Double-voicing at Work – Power, Gender and Linguistic Expertise*, London, Palgrave Macmillan.

Billig, M. 2005. *Laughter and Ridicule: Towards a Social Critique of Humour*, London, Sage.

Boyd, A. 2002. 'Irony, meme warfare, and the extreme costume ball'. In B. Shepard & R. Hayduk (eds), *From ACT to the WTO: Urban protest and community building in the era of globalization*. Verso, London.

Boyer, D. and Yurchak, A. 2010. American Stiob: Or, What late-socialist aesthetics of parody reveal about contemporary political culture in the west, *Cultural Anthropology*, 25(2), 179–221. https://doi.org/10.1111/j.1548-1360.2010.01056.x

Branagan, M. 2007. The last laugh: humour in community activism, *Community Development Journal*, 42(4), 470–481. https://doi.org/10.1093/cdj/bsm037

Bruner, M.L. 2005. Carnivalesque protest and the humorless state, *Text and Performance Quarterly*, 25(2), 136–155. 10.1080/10462930500122773

Cain, C.L. 2012. Integrating dark humor and compassion: identities and presentations of self in the front and back regions of hospice, *Journal of Contemporary Ethnography*, 41(6), 668–694. https://doi.org/10.1177/0891241612458122

Cassegard, C. 2012. Play and empowerment: the role of alternative space in social movements, *Electronic Journal of Contemporary Japanese Studies*, 12(1). http://www.japanesestudies.org.uk/ejcjs/vol12/iss1/cassegard.html

Cottom, T. 2018. North high school senior want to help other be as death positive as she is, *Akron Beacon Journal*, 20 March. Available at: http://bit.ly/2PtvGPQ

Davies, C. 2007. Humour and protest: jokes under communism, *International Review of Social History*, 52, 291–305. https://doi.org/10.1017/S0020859007003252

Death Positivity Movement Blog. 2016. What's missing?, *Death Positivity Movement Blog*, 12 December. Available at: http://bit.ly/2MTapgG

Death Rookie. 2018. The politics of death and dying: a challenge to the death positive movement, *Death Rookie*, 26 September. Available at: http://bit.ly/2ovgQgC

Doughty, C. 2018. What death positive is not, *The Order of Good Death*, 25 January. Available at: http://bit.ly/2pcZwNR

Etsy store The Good Death: Death Positive Goodies and More. Available at: https://www.etsy.com/shop/TheGoodDeath

Foltyn, J.L. 2010. To die for: Skull style and corpse chic in fashion design, imagery, and branding, *Scan*, 7(2). Retrieved from http://bit.ly/2DMVcbu

Gilmour, P. 2016. Why women are leading the death positive movement, *Dazed*, 11 November. Available at: http://bit.ly/2qTvXkN

Göttke, F. 2015. Burning effigies with Bakhtinian laughter, *European Journal of Humour Research*, 3(2–3), 129–144. http://dx.doi.org/10.7592/EJHR2015.3.2.3.gottke

Hart, M. 2007. Humour and social protest: An introduction, *International Review of Social History*, 52(S15), 1–20. www.jstor.org/stable/26405479

Hart, M. 2016. The role of humor in protest cultures. In *Protest Cultures: A Companion*, Eds K. Fahlenbrach, M. Klimke and J. Scharloth, pp. 198–204, New York, NY, Berghahn Books.

Hoy, M. 1992. Bakhtin and popular culture, *New Literary History*, 23(3), 765–782. http://dx.doi.org/10.2307/469229

Kelly, K. 2017. Welcome the reaper: Caitlin Doughty and the 'death-positivity' movement, *The Guardian*, 27 October. Available at: http://bit.ly/31WElwF

Knickerbocker, C. 1964. Humor with a moral sting, *New York Times Book Review*, 27 September. Available at: https://nyti.ms/2RkPeqk

Kowsar, N. 2012. Being funny is not that funny: Contemporary editorial cartooning in Iran, *Social Research*, 79(1), 117–144. https://www.jstor.org/stable/23350301?seq=1

Kutz-Flamenbaum, R.V. 2014. Humor and social movements, *Sociology Compass*, 8(3), 294–304. https://doi.org/10.1111/soc4.12138

Laundry Bear Games. 2017. *A Mortician's Tale*. Laundry Bear Games

Leland, J. 2018. The death positive movement comes to life, *New York Times*, 22 June. Available at: https://nyti.ms/2Wr70sK

Lofland, L.H. 1978. *The Craft of Dying: The Modern Face of Death*. Thousand Oaks, CA, Sage Publications.

Marwick, A. 2013. *Status Update: Celebrity, Publicity and Branding in the Social Media Age*, New Haven, CT, Yale University Press.

Obrdlik, A.J. 1942. 'Gallows humor' – A sociological phenomenon, *American Journal of Sociology*, 47(5), 709–716. https://www.jstor.org/stable/2769536?seq=1

Ortiz, J. 2016. The women who love death, *Marie Claire*, 31 October. Available at: http://bit.ly/2Wp8nb3

Palmer, J. 1994. *Taking Humour Seriously*, London, Routledge.

Pasquali, E.A. 2003. Humour: an antidote for terrorism, *Journal of Holistic Nursing*, 21(4), 398–414. https://doi.org/10.1177/0898010103258602

Petrović, T. 2018. Political parody and the politics of ambivalence, *Annual Review of Anthropology*, 47(1), 201–216. https://doi.org/10.1146/annurev-anthro-102215-100148

Red Ice Podcast. 2017. Dinesh D. Souza is wrong and the death positive movement, *Red Ice Podcast*, 4 September. Available at: http://bit.ly/33PPn7K

Santa-Maria, C. 2019. How Caitlin Doughty is rethinking death, *SOCAL Connected*. Available at: http://bit.ly/2ONZYf9

Senft, T.M. 2008. *Camgirls: Celebrity and Community in the Age of Social Networks*, New York, NY, Peter Lang Publishers.

Speier, H. 1998. Wit and politics: an essay on power and laughter, *American Journal of Sociology*, 103(5), 1352–1401. https://doi.org/10.1086/231355

Sullivan, P. 2012. *Qualitative Data Analysis Using a Dialogical Approach*, Thousand Oaks, CA, Sage Publications.

Swart, S. 2009. 'The terrible laughter of the Afrikaner' – Towards a social history of humor, *Journal of Social History*, 42(4), 889–917. https://doi.org/10.1353/jsh/42.4.889

Thorson, J.A. 1985. A funny thing happened on the way to the morgue: some thoughts on humor and death, and a taxonomy of the humor associated with death, *Death Studies*, 9(3–4), 201–216, doi:10.1080/07481188508252518

Turner, V. 1974. Liminal to liminoid, in play, flow, and ritual: an essay in comparative symbology, *Rice University Studies*, 60(3), 53–92. https://hdl.handle.net/1911/63159

Watson, K. 2011. Gallows humor in medicine. *Hastings Center Report*, 41(5), 37–45. doi:10.1002/j.1552-146x.2011.tb00139.x

Wear, D., Aultman, J.M., Varley, J.D. and Zarconi, J. 2006. Making fun of patients: medical students' perceptions and use of derogatory and cynical humor in clinical settings, *Academic Medicine*, 81(5), 454–462. doi:10.1097/01.ACM.0000222277.21200.a1

Chapter 12

Some Games You Just Can't Win: Crowdfunded Memorialisation, Grief and *That Dragon, Cancer*

Matt Coward-Gibbs

Abstract

January 2016 saw the final release of Numinous Games' crowdfunded linear adventure game *That Dragon, Cancer*. An impactful independent title which subverts many of gaming's traditional and valued norms. In less than two hours of abstracted adventure, players are transported through a series of vignettes documenting one family's struggle with cancer, and the battle faced by their terminally ill child, Joel. Digital memorialisation has been documented by scholars since the late 1990s. This has come in the form of sites specifically created for memorialisation, social networking sites repurposed by their users for memorialisation (*MySpace* and more recently *Facebook*), and online virtual worlds (*Second Life* and *World of Warcraft*). However, within *That Dragon, Cancer* the productive nature of grief has created and envisioned a gaming experience purpose-built for memorialisation. This chapter begins by documenting memorialisation within virtual environments. From here, the author turns to consider the way in which *That Dragon, Cancer* provides a purpose-built space for grief, memorialisation and understanding, focussing on key stylistic and mechanic-based decisions undertaken in the games design. Finally, the author considers the way in which *That Dragon, Cancer*, through the use of crowdfunding in late 2014, transformed from a project memorialising one child to the memorialisation of many across the globe.

Keywords: Crowdfunding; digital games; memorialisation; experience; death; grief

Death, Culture & Leisure: Playing Dead, 173–188

doi:10.1108/978-1-83909-037-020201019

Introduction

That Dragon, Cancer (Numinous Games, 2016) (subsequently *That Dragon*) is not a game that can be won, only experienced. It presents the story of a family's struggle, hope, grief, pain and suffering. In less than two hours of point and click adventure, embedded micro-games and vivid narrative, players are brought close to the heart of the Green family and the story of their son Joel's battle with a terminal Atypical Teratoid Rhabdoid Tumour.[1] *That Dragon* is more than just a game: it is the culmination of productive grief and a powerful memorial for the thousands touched by both Joel's story and cancer the world over. It is crowdfunded memorialisation. In the words Joel's mother and *That Dragon* co-designer Amy Green (2017), it is a 'story that began with a miracle and ended as a memorial'.

In this chapter, I consider the way in which memorialisation is experienced in digital game play. Akin with the work of Anable (2018, 134), I posit that not only are 'video games … affective archives' but, rather than being a static repository, the game space embeds the key principles of memorial within the wider context of playful, flow-like experience (Csikszentmihalyi, 2002). Online memorials have been considered by a number of scholars as archival spaces (Acker and Brubaker, 2014; Bhattacharya, 2010; Recuber, 2012). Furthermore, Spokes (2018) in his work on *Fallout 4* (Bethesda Game Studios, 2015) considers the way in which players themselves move past the game to form their own secondary memorial archive based on their encounters within the digital environment. As such, through using *That Dragon* as an exemplar, this chapter considers the way in which memorialisation takes place when individuals engage in play within the memorial itself.

In the first section of this chapter, I consider the growing role of digital memorialisation before discussing the specific previous incarnations of memorialisation, primarily within online roleplaying games. From this, I turn to consider the way in which the crowdfunded nature of *That Dragon* allows for the provision of a purpose-build space for experiential grief and memorialisation. Following this, I consider the way in which key stylistic and mechanical choices during the games design heightened the archival nature of *That Dragon* before finally addressing the way in which *That Dragon* moved from the memorialisation of a single child to a both experiential and cathartic outlet for numerous individuals involved in the grieving process.

Memorialisation in Digital Gaming Environments

Memorialisation in digital gaming environments is interwoven with the wider practice of online memorialisation. As Walter (2015, 226) contends, internet

[1]Recent literature has suggested that the rendering of illness as a 'battle' or 'fight' can have negative connotations, leading to some to state that negative cancer metaphors can undermine its treatment and prevention (Hauser & Schwarz, 2019). In this chapter, however, I use both terms 'battle' and 'fight' as this is the way in which they are both engaged with by *That Dragon, Cancer*, its accompanying documentary, and in the variety of interviews with the game's design team.

practices associated with mourning have allowed the process to 'become more social, more shared and potentially more conflictual.' Yet the rise of social media has also 'afford[ed] the dead themselves the possibility of becoming more vibrantly present amongst their network of family and friends' (Walter, 2015, 226). In their earlier considerations, Walter et al. (2011, 295) suggest that the effects of online memorialisation are entirely dependent on the way in which those engaging with it relate to the practices. Those who engage publicly with online vlogging, for example, sit in juxtaposition to a social network member connected to but a handful of their closest friends and family members. What is clear, however, is that social media has carved out a space inside the 'broader ecology of practices surrounding death' (Brubaker et al., 2013, 161). In their work concerning the social media practices of the parents of still-born children on Facebook, Hayman et al. (2018, 69) underpin this notion, suggesting that the social network not only allows for the socialisation of the surviving parent, but also provides the space in which 'the stillborn child [is viewed] as an authentic social member'. Outside of more commonplace forms of social networking (Facebook and Twitter to a lesser extent), blogs and online forums have also been adapted as public purpose-built spaces for memorial. Haverinen's (2014a) work, for instance, points to a continuum of privacy and publicity between both the activity and location of memorial. These spaces are distinctive for their users. When considering the construction of memorial blogs publicly, they become spaces which are not just for the expression of grief but also a space in which they are able 'to give solace and information to others in the same situation' (Haverinen, 2014b, 192).

The rise of digital gaming and virtual worlds has provided a rich environment for digital memorialisation, which has been most readily documented by scholars considering the digital landscapes of *Second Life* and *World of Warcraft*. For over 20 years, digital communication tools have played an ever-growing role in individuals lives, with 'virtual universes' becoming the 'social spaces in which the lives of our contemporaries are played out' (Servais, 2015, 363). Yet, as Haverinen (2014a, 157) suggests, grief and bereavement for a co-inhabitant or player within these gaming environments can be undervalued by non-community members, thereby reducing the considered value of gaming experiences.

Second Life (an online virtual world which launched in 2003, see Boellstorff, 2008; Gibson, 2017; Grieve, 2015 for further information) provides a particularly poignant example of memorialisation due to the near-endless possibilities for creation within its virtual landscape. Tom Boellstroff (2008, 128) recalls attending funerals for avatars who had died in 'real life', noting that 'the death of a person in the actual world was usually interpolated as bringing about the death of any avatars that the person had'. These funerals or memorial services for avatars led to the creation of specific sites for the digital dead. Much as within 'real life' the dead of virtual landscapes can be 'mourned on their own terms' (Gibson, 2017, 236) mirroring the mourning experiences of such a 'real life'. However, as Gibson continues, there are instances in which the real life and the second life become interwoven, showing memorials that reveal 'degrees of RL [real life] inclusion of family and friends within SL [*Second Life*] social networks and activities' (Gibson, 2017, 236).

World of Warcraft, however, presents a different environment for memorialisation to take place in as the virtual environment is not editable by the player. As Haverinen (2014a, 157) contends 'every memorial has to be requested from the Blizzard Company, which is why many of the memorials are either for developers … or other known figures in the community'. In 2014, Robin, a genie non-player character was added into the game's the source code for an expansion to *World of Warcraft*, in honour of the late Robin Williams (Kollar, 2014; Schreier, 2014), following a change.org campaign initiated by the player community (Change.org, 2014). Although players are unable to control the genie, they do have an opportunity to engage in limited interactions with him. Outside of these high-profile cases and close associates of Blizzard's development team, memorialisation within *World of Warcraft* has mainly been focussed around events, rather than the creation of spaces (Gibbs et al., 2012, 2013; Haverinen, 2014a, 2014b; Nardi, 2010; Servais, 2015). In particular, Gibbs et al. (2012, 2013) offer an insight into the difficult boundary work which takes place between worlds 'real' and 'virtual', and between play and memorial. In 2006, whilst the guild (a formal in-game collective of players) Maledictions were mourning the death of one of their members via a memorial event, two opposing guilds (Serenity Now and Gnomeland Security) charged the congregation and slaughtered all mourning player avatars present. These actions led to numerous reflections about 'morally acceptable behaviour in an online game and on the Internet more generally' (Gibbs et al., 2013, 2).

Outside of MMO (massively multiplayer online) gaming environments, there have been other instances in which memorialisation has taken place. Primarily, this is reserved for the developers and their close associates. The market place in *Apex Legends'* (Respawn Entertainment, 2019) King's Canyon map contains a small memorial to Sadie, a developer's dog who died suddenly during the game's development. This is a small location and is easily missed, but contains a dog bed with a small dog toy on top of it, coupled with an actual photograph of Sadie contained in a frame. In the action-adventure shooter *Just Cause 4* (Avalanche Studios, 2018), players can locate and navigate to a mountain containing a number of grave markers documenting and memorialising the losses experienced by the development team throughout the game's creation. As the above section has demonstrated, memorialisation does actively take place within digital gaming environments. *That Dragon* is, however, not a game that contains elements of memorialisation: it is a memorial. The playable memorial nature of the game is further compounded when considered alongside the crowdfunding process enacted by the development team which made the game viable.

Crowdfunding Platforms for Game Development

Although *That Dragon* is a closed gaming experience with no multiplayer possibilities, many individuals have been drawn together by the game through its use of the crowdfunding platform *Kickstarter*. Crowdfunding still proves resistant to definition within academic circles, with earlier writers tending to describe it vaguely as 'the financing of a project or venture by a group of individuals instead of professional parties' (Schwienbacher and Larralde, 2012, 370).

More recently, however, this definition has been extended, noting that crowdfunding practices speak to 'an entrepreneur's direct solicitation (often through internet platforms such as Kickstarter or Indiegogo) to a large number of individuals (i.e., "the crowd")' (Short et al., 2017, 149). Whereas previous incarnations of crowdfunding approached the financing of a project as either the lending of monies or viewed the funders as company investors, most game-based projects crowdfunded via online platforms adopt a reward-based model of funding. This means that individuals partaking in the funding receive some form of recompense for their monetary contributions (Agrawal et al., 2011, 2013; Booth, 2015; Kim et al., 2015; Mollick, 2014; Schwienbacher and Larralde, 2012). These rewards could be as simple as a name being displayed on a website, a copy of the final product, a likeness being incorporated into the game, or some form of production credits being given to the funder.

Crowdfunding has proved appealing to independent game developers for a variety of reasons, including the unconstrained nature of funding geographies, meaning that potential funders from across the globe are able to engage in its financing (Agrawal et al., 2013, 4). This is especially true in the case of the crowdfunding of digital games, in which the final product can be delivered to backers in the form of an online key for a digital distribution platform such as Valve's *Steam* or Epic's *Game Store*. The link between crowdfunding and memorialisation within *That Dragon* is stark. Alongside several possible rewards, patrons actively had the opportunity to embed their own memorial within the game.

Numinous Games began development of *That Dragon* in November 2012. Two years later, the project was launched on the largest crowdfunding platform, *Kickstarter*. During a period of 28 days, 3,687 investors pledged a total of $104,491, surpassing the funding goal of $85,000 by over 20% (Numinous Games, 2014). Pledges for 'patrons' in the campaign ranged from the $15.00 digital download code to single pledge of $4,000 to have the patron's 'favourite dog' encoded into the game for Joel to interact with (Numinous Games, 2014). During this process, 58 individuals chose to become a 'patron of the arts' (at a cost of $200) to have an image imported into the game. These images, according to the given specifications could have been to 'commemorate a loved one, represent a memory, be a child's drawing, or just be a beautiful image' (Numinous Games, 2014) (images imported into the game can be seen in the background of Fig. 1). At the $500 tier 'patron publisher' a piece of text up to 250 words was embedded into the game in the form of a message in a bottle. Although only nine patrons chose to back at this tier, their messages (discussed later in the chapter) are poignant reflections of loss, memorialisation and vividly document the multi-faceted presentation of grief within the game.

Embedded memorialisation, however, was not restricted to two tiers but was also available as an optional addon for all pledge levels at a fixed cost of $25.00. This option, known as the 'memorial add-on pledge' allowed backers to add either an 'ink based hand or footprint' to one of the walls of a scene or add a short fifty word message (Numinous Games, 2014). Within *That Dragon* players are playing with the memorial itself, and in this play, they are both reminiscing, engaging with grief and memorialisation.

Mechanisms of Experience

As Jesper Jull (2011, 12–13) contends, 'most video games are ruled *and* make-believe' and in their contexts provide a separate space of enquiry and of opportunity. Huizinga (1949, 10) refers to this space just six times in the construction of his formative work on play as the 'magic circle': a voluntarily entered, liminal and ritualised environment (Rodriguez, 2006). However, it is necessary here to draw attention to the work of Linser et al. (2008) whose adoption of a single word transforms the social-scientific understanding of gaming environments, whether multiplayer or singular, analogue or digital: that word is *porous*. The use of that single word allows entry to a space of play that is not separate from the activity at hand, meaning that a player is able to both feed into and take from gaming environments automatically and willingly. It is this porous nature of the magic circle that allows for *That Dragon* to be 'both a highly personal, yet universal story' (Schott, 2017, 8). Crawford (2015, 577) contends that 'focusing on play in an isolated space, *centres* it, and runs the risk of ignoring the wider social context'. Experiences in *That Dragon* are not separate but form a continuation of social engagement woven throughout the game itself, its *Kickstarter* campaign, and the player's wider social experiences that are informed though its consumption. Muriel and Crawford (2015, 93) speak to the embodied experiential nature of video games, noting that the 'video gamer is always essential to the game experience'. It is this sensorial nature that is embedded throughout *That Dragon* that allows individuals to play within the memorial and play with the memorialised, underpinning the affective archival nature of games (Anable, 2018, 133–134).

Whereas the majority of recent highly-valued (AAA) games, the likes of *Horizon: Zero Dawn* (Guerrilla Games, 2017) or *Assassin's Creed: Origins* (Ubisoft Montreal, 2017) have offered 'the user the possibility to choose alternative routes within a narrative' (Vosmer and Schouten, 2014, 141) the intentional choice by developers to subvert these norms in *That Dragon* offers players the opportunity to 'discover for themselves that there is nothing that they can do that will change the outcome for Joel' (Amy Green, 2017). In rendering the player powerless within the gaming environment, *That Dragon* relegates the active player to the passive role of witness by providing little to no autonomy over their own actions. Time becomes the only way in which the player can seek to regain control over their actions. As Schott (2017, 8) comments, the game instils a 'desire to dwell in its space, to cherish the moments you spend with Joel, and fend off the inevitable'. Time is lucid throughout *That Dragon*. Its vignettes are untamed with minutes eked out and years flashing by in front of the player's eyes. In this sense, *That Dragon* resembles sports-themed games that contract full-length football games into the space of but a few minutes to aid the games flow (Crawford, 2015, 586).

Scene four, 'End of Treatment Party' provides a poignant example of the developers' use of embedded mechanics to demonstrate the passage of time. In a racing style simulation akin to *Super Mario Kart* (Nintendo EAD, 1992), the player guides Joel and his Mother around three short laps of the hospital floor, collecting presumed prizes and using the speed-boosting floor tiles. The timer that is on the top of the screen is not counting your race time but racing through

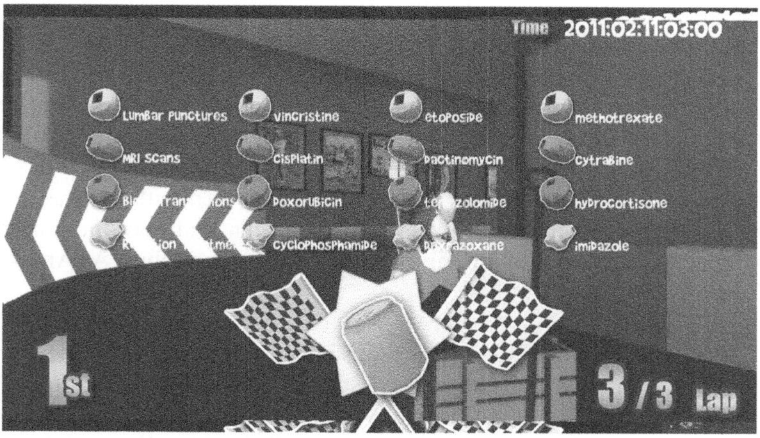

Fig. 1. Screen capture from "End of Treatment Party" taken from *That Dragon, Cancer*.

the thirteen months from January 2010 until February 2011 which Joel spent inside a specialist treatment centre. Crossing the finish line, the player finds that the items they rushed to collect are not bonuses, but account for Joel's regimen of treatment undergone during this period, including: lumbar punctures, blood transfusions, radiation treatments and MRI scans (see Fig. 1). Here, during this embedded game, agency is once again stripped from the player, whose only ability is to decide how long they choose to spend with Joel; how long they wish to spend in the moment. With the race, time becomes limited, constricted to but three laps with no opportunity for a rerun. As Tanz (2016) reflects,

> [...] the overwhelming sensation is one of being a bug caught in a rushing river; you might veer a few degrees in either direction, but you can't alter the flow.

During play, individuals are engaged with the archive. However, the archive and the memorial interact within *That Dragon*. Players are a witness to Joel's life.

Other instances greatly extend periods of time. In scene seven, 'Sorry Guys, It's Not Good', players find themselves in a doctor's office interacting with Joel by playing with a see 'n' say. It is during this scene that Ryan and Amy Green, Joel's parents, are given the news that their son's cancer is terminal. The player uses the see 'n' say to direct the conversation, reversing time to hear the different overlays and voiceovers of his medical care team and his parents. Water begins to rain down into the office, slowly submerging all within it. Again, the player is able to return to the see 'n' say in order to rewind time, stay in the moment, hear what is said again, and attempt to process the given information. Ultimately, the office will always flood at the end of the scene and the player, alongside Joel, will be cast afloat in a sea. These emotive moments are tense. As such, *That Dragon* is a multifaceted memorial process in the sense that the 'memory is reserved' and that being memorialised is 'worthy to be remembered' (OED Online, 2019).

The design of the embedded games within *That Dragon* provides the player with the understanding that they are unable to change Joel's diagnosis. Other embedded games also present spaces wherein the hopelessness of the family's situation can be experienced. Scene nine, 'Joel the Baby Knight', transports players into an 8-bit arcade gaming experience, in which the player controls Joel's traversal through an alternative gaming environment as he battles monsters, runs, hides, and is eventually found by a Dragon, which, even with assistance, he is unable to defeat. This vignette is pivotal to the overall narrative of the game and a poignant act of memorialisation for the Green family themselves. As Amy Green (2017) recalls:

> I cuddled up with my two older sons in bed … I never really knew how much they understood, so I started telling them a bedtime story. I told them about this very brave Knight named Joel and his adventure fighting a terrible dragon called cancer. Every night I told them more of the story, but I never let the story end. I was just building up a context that they could understand.

Whilst controlling Joel in this arcade simulation the player will see the digitised half-reflection of Joel's brothers as they lay being told the story by Amy Green (see Fig. 2). She never did have to finish her children's bedtime. As they grew up, they came to understand Joel's diagnosis (Amy Green 2017).

Other smaller periods of activity and game-based experience also attempt to embed this understanding that the outcome for Joel cannot be changed. Prior to Joel's terminal diagnosis, players control him flying, transported by surgical gloves blown up as if balloons. The balloons are struck as he is chased by the black pulsating cancerous cells; however, no matter how skilled the player is, or how hard the player attempts to steer Joel into a safe path, all the balloons will eventually burst, causing him to fall to the ground.

Fig. 2. Screen capture from "Joel the Baby Knight" taken from *That Dragon, Cancer*.

Outside of the embedded micro-games of *That Dragon,* audio too acts as a pivotal experiential mechanism within the game. Most game releases offer audio options to allow players to alter and personalise audio levels – increasing the dialogue, decreasing sound effects, etc. *That Dragon* offers no such mixing. By not allowing the players to alter the in-game volume levels players are offered the experience intended by its designers. Players experience the thundering cries of an inconsolable child at their full volume which are coupled with a vivid and detailed soundscape. As designer Ryan Green recalls in an interview, 'Jon Hillman is a very talented musician. His ability to create very immersive soundscapes really brings the artwork in the game alive' (Akl, 2016). Here, again, we are returned to the fact that *That Dragon* is a space for the Green family's own memorial. The laughter and aural inflections heard throughout the game are Joel's, which were recorded prior to his death. What the player will not hear, however, are Joel's cries and screams. Although the non-player character of Joel is heard crying and screaming at various points of the game, audio samples used within the game are depository recordings. This is a key decision when using a virtual environment as a space of memorial, especially for the immediate family, friends and associates of the Green family. As Tanz (2016) recalls,

> The cathedral scene is over, but Ryan does not offer you the controller back and you do not ask for it .... The room fills with laughter – his actual laughter, recorded before he died. Ryan pushed a few buttons and makes Joel laugh harder .... Take him [Ryan] in, watch him there, crying and smiling, playing with his creation on the other side of the screen.

Crowdfunded Memorialisation

That Dragon is not the first game to explore death. Recent productions such as *A Mortician's Tale* (Laundry Bear Games, 2017; see Chapters 2 and 11 of this volume), *Life is Strange: Before the Storm* (Deck Nine, 2017), *What Remains of Edith Finch* (Giant Sparrow, 2017; see Chapter 7 of this volume) and *Graveyard Keeper* (Lazy Bear Games, 2018) have shown that conversations concerning the presentation of death, disposal and memorialisation are becoming more prevalent within video gaming. Death within gaming environments is no longer viewed purely as a punishment for failure: it is more complex, more nuanced. *That Dragon* sits in the realm of non-meritocratic games that Christopher Paul (2018, 163) discusses in his recent volume, wherein he suggests that in essence, meritocracy is not a 'bad' thing in games, so long as they do not make up the entirely of gaming culture. In essence, Paul's (2018, 163) call to 'offer an alternative kind of experience and, potentially, target a new kind of player' is answered, in part, by *That Dragon* breaking away from 'meritocratic norms in both narrative and design' (Paul, 2018, 166–167).

That Dragon is not unique in its position as a crowdfunded game. What is unique is its approach to crowdfunding by offering a virtual landscape in which its funders were able to engage in play-based memorial. Several scenes throughout

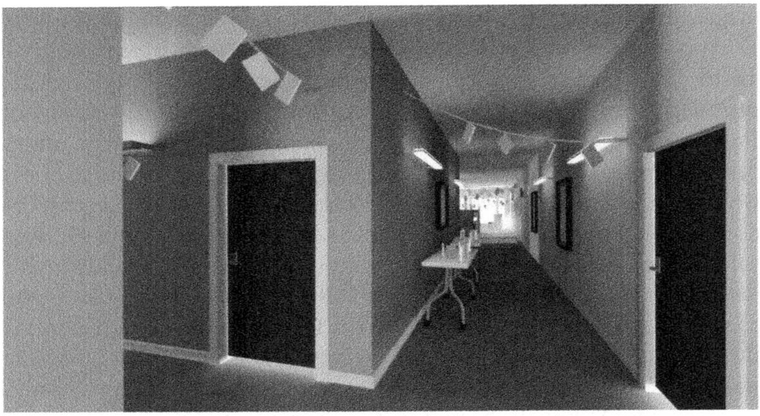

Fig. 3. Screen capture from "waking up" taken from *That Dragon, Cancer*.

the game provide spaces for these communal memorials. Within such spaces, players are invited to read notes (both long and short), view paintings and see pictures memorials of those who have both fought and survived or lost their battle with cancer and other terminal illnesses. The majority of these are from the corpus of 'memorial add on pledges' (see earlier discussion) in which backers were able to provide either a handprint or a short message of up to fifty words (see Fig. 3).

These short messages are found in scene six, 'waking up', which begins with Joel cradled in his mother's arms. A couple of cards sit on the table, the wall to the left displaying several hand and foot prints. As the player slowly turns around in the space, they notice more cards covering all available surfaces – desks, tables, chairs – strung like bunting throughout the hospital rooms and throughout the halls. The vivid colour of the cards makes them all consuming as they litter each corner of the player's view. The messages themselves are considerably varied. Some act as a minimal memorial featuring a name coupled with a date of birth and a date of death, akin to a grave marker or a small obituary in a local newspaper. Others are messages of thanks to hospitals and specific care units. There are also notes speaking of strength addressing the wider population of people with cancer or life limiting illnesses. There are also those which are more cryptic in nature, meant only to be read and understood by the individual writer and their close associates. Many of these messages contain similar motifs, speaking to Christian themes of the afterlife, documenting the strength and value of those being memorialised, and of disease as an adversary to be fought. These cards provide the most direct approach to collective memorialisation within *That Dragon*.

In scene eight, 'Adrift', just following Joel's terminal diagnosis, players can interact with the longer messages supplied by those who pledged as 'patron publishers' at $500. These spaces offer opportunities for longer considerations, which are primarily given in the form of personal reflections. One which stands out is a multi-part correspondence between a game designer and a mother who recently lost her daughter to cancer. The mother writes to the designer to tell him of the joy his game brought her daughter in the latter part of her illness and how it lifted her spirits. Too, we also see the designer's response, offering condolences and including a typescript of an in-game adventure that her daughter had been on stating that she was a member of the Knight's guild and had 'defended the city many times from the forces of evil'. He continues, writing that 'you[r] daughter may have ultimately lost her battle with that dragon, cancer, but in my game, she was the dragon slayer'. There is, as such, a continuum of memorial narratives throughout *That Dragon* from the simplest of handprint, to the emotive reflections of a parent on the loss of their child. *That Dragon*, though, is more than just a sensory archive. It is a playful archive. Though, being a playful memorial is not without its own contentions.

Games Should be Fun

In an address at SDF Ryan Green (2016) considered the design process of *That Dragon*. He reflects on the criticism that the game has garnered, stating:

> If this act of memorial seems odd to you or maybe even in poor taste, you're not alone in your reservations. Often criticism of our choice to memorialise our son in the way takes the form of 'cancer's not a game' or 'you should have made a film'... 'games should be fun' or 'who's going to play that'.

He continues his reflection, later contending that it was through play that he and Joel connected. As such, a game became the perfect catalyst for the telling of Joel's story. Through the mechanical and stylistic decisions during the design process such as the removal of player autonomy and lucid approaches to time, *That Dragon* creates spaces for reflection and spaces in which, through play, the archive can be further unpicked. The game, being constructed around the story of Joel's short life, means that it never can be truly centred, for it is always enveloped in its wider social significance (Crawford, 2015; Muriel and Crawford, 2018).

The ethic of *That Dragon* is something which needs reading against the wider literature. In her consideration of storytelling in palliative care, Rattner (2019) builds on the autoethnographic principle of 'vulnerability with a purpose'. She argues that un-disciplining the discourses around mortality allow for 'untellable stories' to become 'more tellable' (Rattner, 2019, 363).

Unsurprisingly, following the game's release, it was picked up by several influential YouTubers. Although clickbait in nature, the titles of their videos such as 'emotionally drained' (jacksepticeye, 2016), 'this game will make you weep' (PewDiePie, 2016), 'all aboard the feels train' (Mini Ladd, 2016) and 'misery simulator' (Gloom, 2016) hint at the wider social and emotional engagement with the narrative of *That Dragon*. Seán McLoughlin (Jacksepticeye) who is primarily known for his high-energy commentaries, for example, spends time during play exploring his grief at the loss of his grandmother. He reads the cards littered throughout the landscapes and speaks directly to Joel, elongating the time spent with him. He reaches the end of the game, not knowing that the story he had just

Fig. 4. Screen capture from "picnic at the edge of the world" taken from *That Dragon, Cancer*.

enacted was reality. As the game fades to white the words 'thank you for playing' appear for a moment, before he is greeted with an image of Joel. 'Oh, he was real', McLoughlin exclaims, he takes a moment to collect himself before continuing: 'I was really hoping all that time that it wasn't real and that it was just a story'. He engages with each image as the credits roll. In essence, McLoughlin's engagement with the story of *That Dragon* figured Joel as the authentic social member, much the same as Hayman et al. (2018) noted with reference to the parasocialisation of stillborn children on Facebook.

Thank you for playing is an important statement here as *That Dragon* is not attempting to fetishise terminal illness or make light of the tumultuous experiences associated with it. The Green family and the wider design team are very much aware the overarching moniker that 'cancer is not a game'. *That Dragon* provides a space to memorialise and reflect, a chance to hear their son's laughter. As Sicart (2014, 59) contends, spaces of play can be defined as 'giving us a place to be'. It is rudimentary in a sense, but considering the play space as a place of reflexive liminality allows for a true decentring of play and an open engagement with the affective archive (Crawford, 2015; Anable, 2018).

Concluding Remarks

That Dragon is the result of productive grief and crowdfunded memorialisation. However, engagement with the game is a form of playful memorial. On the surface, it is the story of Joel Green, but scratching at the surface reveals a multitude of interwoven memorial narratives, each of which play out within Joel's story. The black twisting and pulsating form of disease are present throughout *That Dragon*. It lingers in the experience of its players and is ever-present throughout the game until its epilogue: 'picnic at the edge of the world' where it can no longer be found. Joel has died, but the player can still experience his laugh, his love of bubbles and of dogs. In this space, he speaks for the first time and is happy to see the player, who has spent much of this journey by his side, once more. He sits eating pancakes with a dog called Manju, who was embedded into the game as a result of the crowdfunding process, again exemplifying the way in which the game intersperses narratives.

At several points throughout this game, the player is offered these 'places to be' (Sicart, 2014, 59). The dragon's form, however, is not forgotten. It lingers in both Joel's memorial and his parents' grief (Fig. 4). The Green family's engagement in productive grief has not only extended into the creation of *That Dragon* but also in the development of its accompanying documentary *Thank You for Playing* (Osit and Zouhali-Worrall, 2015) and the Children's picture book *He's Not Dead Yet* (Green and Green, 2013; see Chapter 4 of this volume, for a further discussion of the presentation of death within children's picture books). Each of these tools considered in this chapter offer spaces for conversations and contemplations of death which can be introduced openly and expressively. Death is emotive, more so it brings about a multitude of sometimes conflicting emotional states, which too are captured throughout the game. *That Dragon* is as much about the people who play, as it is about the Green family's grief, and Joel.

Interactions and Sensoria

Dewey, S. 2015. *Ten Candles: A Storytelling Game of Tragic Horror*, Cavalry Games.
Fox, M. and O'Connor, R. 2018. *Holding On: The Troubled Life of Billy Kerr*, Hub Games.
Lazy Bear Games. 2018. *Graveyard Keeper*, Tiny Build.
Osit, D. and Zouhali-Worrall, M. (Director). 2015. *Thank You for Playing*, Kinematic Films (80 minutes).

References

Acker, A. and Brubaker, J.R. 2014. Death, memorialization, and social media: A platform perspective for personal archives, *Archivaria: The Journal of the Association of Canadian Archivists*, 77, 1–23.
Agrawal, A., Catalini, C. and Goldfarb, A. 2011. *The Geography of Crowdfunding*. Working Paper No. 16820. Cambridge, MA, National Bureau of Economic Research.
Agrawal, A., Catalini, C. and Goldfarb, A. 2013. Some simple economics of crowdfunding, *Innovation Policy and the Economy*, 14, 63–97. http://dx.doi.org/10.1086/674021
Akl, A. 2016. 'That dragon, cancer': Q&A with developer Ryan Green, *Voice of America*, 22 January. Available at: http://bit.ly/2MZ7Olz
Anable, A. 2018. *Playing with Feelings: Video Games and Affect*, Minneapolis, MN, University of Minnesota Press.
Avalanche Studios. 2018. *Just Cause 4*, Square Enix.
Bethesda Game Studios. 2015. *Fallout 4*, Bethesda Softworks.
Bhattacharya, S. 2010. Mourning becomes electronic(a): 9/11 online, *Journal of Creative Communications*, 5(1), 63–74. https://doi.org/10.1177/097325861100500104
Boellstorff, T. 2008. *Coming of Age in Second Life: An Anthropologist Explores the Virtually Human*, Princeton, NJ, Princeton University Press.
Booth, P. 2015. Crowdfunding: a spimatic application of digital fandom, *New Media & Society*, 17(2), 149–166. https://doi.org/10.1177/1461444814558907
Brubaker, J.R., Hayes, G.R. and Dourish, P. 2013. Beyond the grave: Facebook as a site for the expansion of death and mourning, *The Information Society*, 29(3), 152–163. https://doi.org/10.1080/01972243.2013.777300

Change.org. 2014. Sign the petition: create an NPC within world of warcraft to memorialize the Late Robin Williams, an avid world of warcraft player, and one of the funniest men alive. *Change.org*. Available at: http://bit.ly/2NnaIPW

Crawford, G. 2015. Is it in the game? Reconsidering play spaces, game definitions, theming, and sports videogames, *Games and Culture*, 10(6), 571–592. https://doi.org/10.1177/1555412014566235

Csikszentmihalyi, M. 2002. *Flow: The Classic Work on How to Achieve Happiness* (revised ed.), New York, NY, Rider.

Deck Nine. 2017. *Life Is Strange: Before the Storm*, Square Enix.

Giant Sparrow. 2017. *What Remains of Edith Finch*, Annapurna Interactive.

Gibbs, M., Mori, J., Arnold, M. and Kohn, T. 2012. Tombstones, uncanny monuments and epic quests: memorials in world of warcraft, *Game Studies*, 12(1).

Gibbs, M., Carter, M., Arnold, M. and Nansen, B. 2013. Serenity now bombs a world of warcraft online funeral: negotiating the morality, reality and taste of online gaming practices, *Selected Papers of Internet Research*, 14.

Gibson, M. 2017. Grievable lives: Avatars, memorials, and family 'plots' in second life. *Mortality*, 22(3), 224–39. https://doi.org/10.1080/13576275.2016.1263941

Gloom. 2016. Misery simulator – That dragon, cancer. *YouTube*, 11 October. Available at: http://bit.ly/31WtMd0

Green, A. 2017. A video game to cope with grief. *TED NYC*, February. Available at: http://bit.ly/3498rOY

Green, A. and Green, R. 2013. *He's Not Dead Yet*, Media Greenhouse.

Green, R. 2016. Designing for intimacy: That dragon, Cancer – Ryan green | SDF2016, *YouTube*, 4 July. Available at: http://bit.ly/2WlmePV

Grieve, G.P. 2015. The middle way method: a Buddhist-informed ethnography of the virtual world of second life. In *Buddhism, the Internet, and Digital Media: The Pixel in the Lotus*, Eds G.P. Grieve and D. Veidlinger, pp. 23–39, New York, NY, Routledge.

Guerrilla Games. 2017. *Horizon Zero Dawn*, Sony Interactive Entertainment.

Hauser, D.J. and Schwarz, N. 2019. The war on prevention II: battle metaphors undermine cancer treatment and prevention and do not increase vigilance, *Health Communication*, 1–7 September. https://doi.org/10.1080/10410236.2019.1663465

Haverinen, A. 2014a. In-game and out-of-game mourning: on the complexity of grief in virtual worlds. In *Mediating and Remediating Death*, Eds D.R. Christensen and K. Sandvik, pp. 155–174, Surrey, Ashgate.

Haverinen, A. 2014b. *Memoria Virtualis – Death and Mourning Rituals in Online Environments*. PhD thesis, Finland, University of Turku.

Hayman, J., Chamberlain, K. and Hopner, V. (2018). Disembodied social life: the ongoing social presence of the born-still on Facebook, *Mortality*, 23(1), 65–81. https://doi.org/10.1080/13576275.2017.1304367

Huizinga, J.. 1949. *Homo Ludens: A Study of the Play-element in Culture*, London, Routledge & Keegan Paul.

jacksepticeye. 2016. Emotionally drained | That Dragon, Cancer, *YouTube*, 15 January. Available at: http://bit.ly/2MTjNAR

Jull, J. 2011. *Half-real: Video Games between Real Rules and Fictional Worlds*. Cambridge MA, The MIT Press.

Kim, J.-H., Newberry, P. and Qui, C. 2015. An Empirical Analysis of a Crowdfunding Platform. Net Institute Working Paper no. #15-12.

Kollar, P. 2014. Robin Williams memorial character discovered in world of warcraft: warlords of Draenor beta. *Polygon*, 4 September. Available at: http://bit.ly/2MWWeaz

Laundry Bear Games. 2017. *A Mortician's Tale*, Laundry Bear Games.

Lazy Bear Games. 2018. *Graveyard Keeper*, Tiny Build.

Linser, F., Ree-Lindstad, N. and Vold, T. 2008. The magic circle: game design principles and online role-play simulations. In *ED-MEDIA 2008*, Vienna, Austria.
Mini Ladd. 2016. All aboard the feels train! – That Dragon Cancer, *YouTube*, 7 April. Available at: http://bit.ly/2Wo4lQl
Mollick, E. 2014. The dynamics of crowdfunding: an exploratory study, *Journal of Business Venturing*, 29(1), 1–16. https://doi.org/10.1016/j.jbusvent.2013.06.005
Muriel, D. and Crawford, G. 2018. *Video Games as Culture: Considering the Role and Importance of Video Games in Contemporary Society*, London, Routledge.
Nardi, B.A. 2010. *My Life as a Night Elf priest: An Anthropological Account of World of Warcraft*, Ann Arbor, MI, University of Michigan Press.
Nintendo EAD. 1992. *Super Mario Kart*, Nintendo.
Numinous Games. 2014. *That Dragon, Cancer*. Kickstarter. Available at: http://bit.ly/2MTL8TA
Numinous Games. 2016. *That Dragon, Cancer*, Numinous Games.
OED Online. 2019. Memorial, Adj. and n. In *OED Online*. Oxford University Press. Available at: https://www.oed.com/view/Entry/116351
Osit, D. and Zouhali-Worrall, M. (Director). 2015. *Thank You for Playing*, Kinematic Films (80 minutes).
Paul, C.A. 2018. *The Toxic Meritocracy of Video Games: Why Gaming Culture Is the Worst*, Minneapolis, MN, University of Minnesota Press.
PewDiePie. 2016. This game will make you weep (That, Dragon Cancer). *YouTube*, 29 March. Available at: http://bit.ly/36dq6H7
Rattner, M. 2019. Tellable and untellable stories in suffering and palliative care, *Mortality*, 24(3), 357–368. https://doi.org/10.1080/13576275.2018.1530206
Recuber, T. 2012. The prosumption of commemoration: disasters, digital memory banks, and online collective memory, *American Behavioral Scientist*, 56(4), 531–549. https://doi.org/10.1177/0002764211429364
Respawn Entertainment. 2019. *Apex Legends*, Electronic Arts.
Rodriguez, H. 2006. The playful and the serious: an approximation to Huizinga's homo ludens, *Game Studies*, *6*(1).
Schott, G. 2017. That dragon, cancer: Contemplating life and death in a medium that has frequently trivialized both. In *Proceedings of DiGRA 2017*, The University of Melbourne, Melbourne.
Schreier, J. 2014. World of warcraft's Robin Williams tribute is just great. *Kotaku*, 9 April. Available at: http://bit.ly/2PuuATK
Schwienbacher, A. and Larralde, B. 2012. Alternative types of entrepreneurial finance. In *The Oxford Handbook of Entrepreneurial Finance*, Ed. D. Cumming, pp. 369–391, Oxford, Oxford University Press.
Servais, O. 2015. Funerals in the 'world of warcraft': religion, polemic, and styles of play in a videogame universe, *Social Compass*, 62(3), 362–378. https://doi.org/10.1177/0037768615587840
Short, J.C., Ketchen, D.J., Mckenny, A.F., Allison, T.H. and Duane, R. 2017. Research on crowdfunding: reviewing the (very recent) past and celebrating the present. *Entrepreneurship Theory and Practice*, 41(2), 149–160. https://doi.org/10.1111/etap.12270
Sicart, M. 2014. *Play Matters*, Cambridge, MA, The MIT Press.
Spokes, M. 2018. 'War … war Never Changes': exploring explicit and implicit encounters with death in a post-apocalyptic gameworld, *Mortality*, 23(2), 135–150. https://doi.org/10.1080/13576275.2017.1319348
Tanz, J. 2016. A father, a dying son, and the quest to make the most profound videogame ever. *WIRED*, January. Available at: http://bit.ly/36oVBhE
Ubisoft Montreal. 2017. *Assassin's Creed: Origins*, Ubisoft.

Vosmer, M. and Schouten, B. 2014. Interactive cinema: Engagement and interaction. In *Interactive Storytelling: 7th International Conference on Interactive Digital Storytelling*, Eds A. Mitchell, C. Fernández-Vara and D. Thue, Lecture Notes in Computer Science, Vol. 8832, New York, NY, Springer.

Walter, T. 2015. Communication media and the dead: from the stone age to Facebook, *Mortality*, 20(3), 215–232. https://doi.org/10.1080/13576275.2014.993598

Walter, T., Hourizi, R., Moncur, W. and Pitsilldes, S. 2011. Does the internet change how we die and mourn? Overview and analysis, *OMEGA*, 64(4), 275–302. https://doi.org/10.2190%2FOM.64.4.a

Chapter 13

Suicide, Angst, and Popular Music

Christopher Partridge

Abstract

Throughout history, from ballads to requiems, music has played an important social role in reflection on mortality. Not only do musicians articulate their angst and thoughts about death, but, in so doing, they enable listeners to explore their own feelings. While the relationship between music and mental health can be examined from a number of perspectives, two broad approaches can be taken: *artist-centred approaches* and *listener-centred approaches*. The first analyses the life and work of artists, focussing particularly on the ways in which they explore death and angst in their music. The second looks at the ways in which the life and work of an artist is interpreted by listeners. Within these general approaches, a complex set of questions emerge – often at the interface of both approaches. How is the music used by listeners in their reflection on mortality? How is music used to manage mental health? Does reflection on the life and work of an artist contribute to suicidal ideation? Is the reception of music altered by an artist's suicide? Using both these approaches and drawing particularly on the work of Émile Durkheim, this discussion demonstrates the significance of popular music analysis for death studies, focussing particularly on the issues surrounding popular music's relationship to suicidal ideation.

Keywords: Popular music; suicide; angst; depression; Émile Durkheim; Werther effect

Introduction

Ian Curtis, the troubled lyricist and lead singer of Joy Division hung himself on 18 May 1980 shortly after the album *Closer* was recorded and two months prior to its release on 18 July. As soon as the band's grieving and shocked fans placed

Death, Culture & Leisure: Playing Dead, 189–208

doi:10.1108/978-1-83909-037-020201020

the needle on the record, it was clear to them that, effectively, they were listening to a suicide note. *Closer* documents the thoughts of an unhappy and conflicted man. While his 'professional life was flourishing', noted his widow, his 'personal life was disintegrating'. He 'seemed to be in a trance for the whole of the time he was writing and recording the lyrics for *Closer*. Wound up and intense, he was in another world' (Curtis, 1995, 110–111). Having said that, Joy Division fans were used to Curtis's morbidity. Reading through all his published work, it is clear that he struggled with angst.

Curtis was, of course, not the only artist to have been drawn 'back to black', as Amy Winehouse put it (*Back to Black*, 2006).

> Wine is fine
> But whiskey's quicker
> Suicide is slow with liquor
>
> Take a bottle, drown your sorrows
> Then it floods away tomorrows.

These are the opening lyrics to Ozzy Osbourne's 'Suicide Solution' from his 1980 debut album *Blizzard of Oz*. While there has been some disagreement between Osbourne and his co-writer, Bob Daisley, regarding whether this is a song about Osbourne's own self-destructive struggle with alcohol or a personal reflection on the death of his friend, Bon Scott of AC/DC who had recently been found dead after a heavy drinking session (Osbourne, 2009, 290–291), it has become notorious because of another death. On 26 October 1984, a 19 year old, John Daniel McCollum, shot and killed himself while he was lying on his bed, wearing his headphones, and listening to 'Suicide Solution' at his home in California. His parents unsuccessfully sued Osbourne on the grounds that the song was a 'proximate cause' for their son's death. They are, of course, not the only parents to have worried about the effect of popular music on the mental health of their children. Indeed, there have been several failed court cases, including one in which the parents of Raymond Belknap and James Vance claimed that the band Judas Priest had encouraged their sons to commit suicide using subliminal messages on their 1978 album *Stained Class*. Again, following the suicide of Hannah Bond, a fan of My Chemical Romance, the coroner not only referred to their music as disturbing, but suggested that it was implicated in her death. Despite the band's condolences and their insistence on being 'anti-suicide' – 'one of our missions [is] to provide comfort, support, and solace to our fans' (*NME*, 2008) – the suicide was sensationalised in the media. The *Daily Mail*, for example, referred to My Chemical Romance as a 'suicide cult band' (Levy, 2008).

From ballads to requiems, throughout history music has played an important social role in reflection on mortality. Not only do composers and songwriters articulate their angst and thoughts about death, but, in so doing, they enable listeners to explore their own feelings. For example, Lou Reed, following the death of two of his close friends – Rotten Rita (Kenneth Rapp) and the songwriter Doc

Pomus – recorded *Magic and Loss* (Fricke, 1992), a simple and stark meditation on cancer and whether life is worth living. As David Fricke of *Rolling Stone* commented in his review of the album:

> This is an album about death – and how to live with it. It is an eyewitness account, documented in compelling song, of a losing battle with cancer … and the little miracles that, for the mourner, mark the beginning of the healing process …. Anyone who's been through this experience will know exactly what Reed is talking about. For anyone facing it for the first time, *Magic and Loss* can help with the starting over. (Fricke, 1992)

Likewise, Peter Silberman of the band Antlers explored the emotional distress caused by cancer on the 2009 concept album *Hospice*, which relates the story of a friendship between a hospice worker and a patient suffering from terminal bone cancer. Similarly, Penelope Trappes recently commented that her 2018 album *Penelope Two* is the result of spending much of the previous year

> being an empath with two dear friends, who lost very close loved ones. One lost nearly her whole family in an accident and the other lost his partner of 23 years, after she gave birth to their third child.

As such the album is an echo of their 'distance, pain, love and infinity' (Malt, 2018).

Again, Alex Rosamilia, the lead guitarist of The Gaslight Anthem, has recently spoken about his new band, Dead Swords, which is both musically and lyrically a darker and more introspective project. 'I talk about being suicidal or being depressed …. That's stuff I've struggled with throughout my 37 years on this planet' (Pearlman, 2019).

While the relationship between music and mental health can be examined from a number of perspectives, two broad approaches might be taken: *artist-centred approaches* and *listener-centred approaches*. The first analyses the life and work of artists, focussing particularly on the ways in which they explore death and angst in their music. The second looks at the ways in which the life and musical compositions of an artist are interpreted by listeners. Obviously, within these general approaches, a complex set of questions emerge – often at the interface of both approaches. How is the music used by listeners in their reflection on mortality? How is music used to manage mental health? Does reflection on the life and work of an artist contribute to angst and suicidal ideation? Is the reception of music altered by an artist's suicide? There is, thankfully, a growing, though still small body of work beginning to explore this important area of both death studies and popular music studies (see Strong and Lebrun, 2015; Partridge, 2015). The following discussion indicates the importance of popular music analysis for death studies, focussing particularly on the relationship between popular music and suicidal ideation.

Context

While music-related suicides are relatively rare, self-harm per se is not uncommon. The 2018 Samaritans'[1] 'Suicide Statistics Report' (Samaritans, 2018) recorded 6,213 suicides in the United Kingdom and Republic of Ireland. Concerning children and young people under the age of 19, the statistics reveal a more complex picture. For example, the most recent survey by NHS Digital, 'Mental Health of Children and Young People in England, 2017' (NHS Digital, 2018) found that one in eight young people have a mental health disorder, the incidence of which rose to one in six for young people aged 17–19. However, while males are more likely to experience mental health disorders than girls up until the age of 11, between 11 and 16 both sexes are equally likely to suffer, and by the time girls reach 17, they are more than twice as likely to have a mental health disorder. Between the ages of 17 and 19 almost one in four females struggle with mental health problems, half of which indicated that they had self-harmed or attempted suicide. Generally, across all age groups, it would appear that

> despite recent advances in the understanding and treatment of self-harm… progress in reducing suicide death rates has been elusive, with no substantive reduction in suicide death rates over the past 60 years. (Ougrin et al., 2015, 97)

Again, a recent study for the Prince's Trust in the United Kingdom found that 9% of all respondents (which equates to around 751,230 young people) agreed with the statement 'I have nothing to live for'. Furthermore, one in three unemployed young people (32%) had contemplated suicide, while one in four (24%) had self-harmed (Sellgren, 2014).

These trends are, of course, not limited to the United Kingdom. For example, according to the American Foundation for Suicide Prevention, suicide is currently the 10th leading cause of death in the United States. A total of 1.3 million adults made at least one suicide attempt and in 2017. Overall, there were 1,300,000 attempts (American Foundation for Suicide Prevention, 2018). More particularly, the rate of young sufferers continues to rise. Those who experienced a 'Major Depressive Episode (MDE) increased from 11.93% to 12.63%' and '62% of youth with MDE received no treatment' (Mental Health America, 2018; see also Sanburn, 2013, 1).

Concerning the relationship between popular music and poor mental health, while the musicians themselves are often older adults, the listeners most keenly affected are typically adolescents and young adults (i.e. the high risk groups discussed above). Obviously, the relationship is a complex one and while a number of theoretical tools can be used to interrogate the issues, such as those provided by Freudian analysis (Partridge, 2015, 100–104; see also, Kernberg, 2009,

[1]Founded in 1953, the Samaritans is a charity in the UK and Ireland that provides support to anyone in emotional distress or at risk of suicide. See https://www.samaritans.org/.

1009–1023), the following discussion draws on the sociological ideas articulated by Émile Durkheim in his ground breaking 1897 work *Le Suicide*.

Durkheim, Suicide, and Popular Music

For Durkheim, because even our most subjective actions are part of a constellation of sociostructural reflexes, suicide can be accounted for largely in terms of coercion by social forces. While those who self-harm have personal reasons for so doing, in the final analysis, suicide is structural, a 'social fact', a product of a particular relationship between the individual and society. Hence, suicides reveal something about the state of the communities in which they take place. While Durkheim's work in this area has been contested (see Pickering and Walford, 2000), nevertheless, it was enormously important, not least in moving the discussion away from a nineteenth century focus on perceived personal and moral failings and towards a more scientific analysis of the statistics. And, even if, as Christian Baudelot and Roger Establet (2008) have shown, aspects of his thesis need revising, it still provides a useful framework for discussions of suicide. Of course, in the final analysis, it has to be acknowledged that, while any sociological analysis can provide some understanding of angst and self-aggression, because the very act of suicide is individual and exceptional, its causes are often difficult to ascertain. This discussion is simply a contribution to the analysis of suicide from the perspective of popular music studies.

Beginning with a statistical analysis of suicide rates, Durkheim found that the risk of suicide in modern societies is inversely proportional to the degree of social integration. The extent to which individuals are subordinate to the social group – in the sense of sharing rituals, conventions, and beliefs – affects their vulnerability to suicide. For example, from the perspective of religion, those who attend worship regularly and engage in the social life of their communities have a high level of integration and, therefore, a lower risk of suicide. The same is true of integration in family life. Consequently, the evidence demonstrates that divorced people are between two and four times more likely to commit suicide than their married peers of the same age and gender (see Stack, 2000, 163–176; see also Kposowa, 2003, 993). Of course, as well as the sense of meaning engendered by social integration, more practically, the group itself provides important support mechanisms for those of its members who struggle with anxiety and depression.

This brings us to Durkheim's identification of types of suicide: altruistic, egoistic, and anomic. Although he also discussed 'fatalistic' suicides in a footnote, he didn't consider them to be a particularly significant contributor to suicide rates. Indeed, he was far more interested in egoistic and anomic suicides – which will be the focus of this discussion.

We begin with altruistic suicides. These are committed by people who experience high levels of integration. They wholeheartedly identify with the group, its identity and its goals. The level of integration is such that it erodes autonomy and individual identity, encouraging members, if necessary, to sacrifice themselves in the service of the group. Individuals commit suicide for that which they value above personal survival. Moreover, the group tends to approve of and celebrate such 'heroic' acts.

The focus on suicide in popular music is, however, predominantly not altruistic. As in, for example, the song 'Depression and Anguish' by Suicidal Tendencies, from their album *Suicidal for Life* (1994), the focus is introspective and tends to explore self-hatred, hopelessness, anger, and depression. This is perhaps unsurprising, in that, as Donna Gaines's (1998) study of 'suburbia's dead-end kids' found, as part of an interest in transgressive music and discourses, many expressed a sense of helplessness and despair, which led to morbidity and, particularly, to a fascination with violence and self-harm. This, Gaines argues, can be partially explained in terms of Durkheim's (1952, 276) fatalistic suicide, which is a consequence of 'excessive regulation', and is typical of 'persons with futures pitilessly blocked and passions violently choked by oppressive discipline'. Such individuals suffer from the perception of a lack of agency and, consequently, feel trapped. The young people Gaines (1998, 252–253) studied had 'lost their ability to dream' and, as such, felt 'fated, doomed'. There was a feeling, according to one young person she spoke to, that 'they were beaten down as far as they could go'. Hence, "lacking confidence in themselves and the world 'out there', they felt trapped. Without a sense of meaningful choices, the only way out *was suicide*" (Gaines, 1998, 253). That said, 'fatalistic suicide is not the whole story' (Gaines, 1998, 253) in that many such suicides need to be understood as more egoistic and anomic.

Egoistic suicide is a consequence of poor social integration. The perception of having no significant ties to a community engenders feelings of being alone and misunderstood. This leads to an unwillingness and often an inability 'to emerge from' oneself. That is to say, as Durkheim (1952, 278–279) put it,

> what is lost in activity is made up for in thought and inner life. In revulsion from its surroundings consciousness becomes self-preoccupied, takes itself as its proper and unique study, and undertakes as its main task self-observation and self-analysis. But by this extreme concentration it merely deepens the chasm separating it from the rest of the universe.

Such feelings of angst are, of course, not unusual in the liminal cultures of youth. Indeed, they are explicitly articulated and even normalised within some popular music subcultures (see Greenwald, 2003). Goth, emo, hardcore, and some heavy metal music and subcultures, for example, explicitly encourage identities constructed around introspection and melancholia. Emo is a particularly good example of this. The term itself, which is an abbreviation of 'emotional' and 'emocore' (a subgenre of hardcore punk), indicates a focus on the exploration of affective states. Early emocore bands such as Rites of Spring, which emerged out of the angry culture of hardcore, effectively stopped shaking their fists at society and turned their anger within.

> I'm going down, going down
> Deeper than inside…
> 'Till there's nothing left to find.
>
> ('Deeper Than Inside,' *Rites of Spring*, 1985)

Increasingly through the 1990s, certain genres of music focussed on evoking dispositions that reflect an exaggerated sense of vulnerability, despair, social dislocation, and suicidal ideation:

> You gotta go down, deep down into the pain
> Let it purge your soul like flesh to a razor blade
> Dig down deep into the pain
> Surrender, surrender, dig into euphoria.
>
> (Steve Vai, 'Down Deep Into the Pain', *Sex and Religion*, 1993)

> There is nothing more for me
> Need the end to set me free…
>
> Death greets me warm
> Now I will just say goodbye.
>
> (Metallica, 'Fade to Black', *Ride the Lightening*, 1984)

This sense of a loss of meaning and belonging is, again, reflected in Gaines's research in the 'teenage wasteland'. It feels, said one young person, as though 'you don't fit in anywhere, there is no place for you: in your family, your school, your town – in the social order' (Gaines, 1998, 253). The problem is, as Durkheim (1952, 279) put it, 'self-absorption is not a good method of attaching oneself to others'. Hence, it might be argued that subcultural discourses can actually encourage introspection, melancholia, and detachment. In other words, there is a question as to whether popular music can draw vulnerable listeners into pathological states of mind that are maintained by the liminal pursuit of subcultural capital. This type of thinking has informed concerns about the dark pop of a number of young artists whose music appeals to adolescents. A good example is the discussion surrounding Billie Eilish, whose songs explore transgressive themes, including angst and suicide. Her song 'Bury a Friend' from her 2019 album *When We All Fall Asleep, Where Do We Go?* has been described as immersion 'into a depth of darkness that's not only treacherous, but unreachable'. This listener then expresses a concern that is shared by a number of commentators:

> What if I was so immersed in her album when I was going through some of the hardest and most confusing things I've ever gone through in my formative years? I would not feel comforted, uplifted, or reached. I would feel pushed further and forcibly into my own darkness. (Moser, 2019)

While this type of response is obviously a worry, recent research has shown that, actually, for some listeners who suffer from depression, not only does such music induce calm, but that it helps them to negotiate their feelings of angst (Yoon et al., 2019). As one fan of My Chemical Romance (MCR) commented,

> whenever I'm feeling depressed I'll listen to MCR. I've read countless internet posts from kids who are adamant that MCR

> prevented them from cutting themselves, shooting up, and even killing themselves. (aNormaNonEmoKid, 2012)

Furthermore, some musicians specifically use their music to help people work through their pain. Alex Rosamilia, for example, explains that he talks about feelings of angst, rather than about self-harm or suicide because 'there should always be hope'. Discussing the song 'Ender' (*Enders,* 2019), he says that

> it's about pulling yourself out of that hole… But my real point for this band is that it's okay to feel however you feel – just make sure you don't let those feelings take you to a place you can't come back from. (Pearlman, 2019)

Again, from a different perspective, as Fricke commented, for anyone facing bereavement for the first time, Lou Reed's '*Magic and Loss* can help with the starting over' (Fricke, 1992). This, of course, raises the question as to the positive therapeutic value of dark music (see below).

Concerning anomic suicides, these are the result of those sudden and unexpected changes in social circumstances that individuals struggle to adjust to. Typically, sufferers experience detachment from regulatory frameworks in society. For example, 'it is a well-known fact', says Durkheim (1952, 241), 'that economic crises have an aggravating effect on the suicidal tendency'. This is supported by recent research into suicide rates during economic recession. For example, austerity in Greece led to a significant rise in male suicides (Antonakakis and Collins, 2014). This is hardly surprising if we accept Durkheim's (1952, 246) central argument that 'no living being can be happy or even exist unless his needs are sufficiently proportioned to his means'. If an individual's desires surpass the capacity to satisfy them, the result is frustration, misery, and an overall weakening of the impulse to live:

> Unlimited desires are insatiable by definition and insatiability is rightly considered a sign of morbidity. Inextinguishable thirst is constantly renewed torture… To pursue a goal which is by definition unattainable, is to condemn oneself to a state of perpetual unhappiness. (Durkheim, 1952, 247–248)

If societies do not regulate these desires in some way, the individual's experience of life is vulnerable to being one of dissatisfaction and disillusionment. Modern capitalist economies are chronically anomic, driven by, as he puts it elsewhere, 'a crass commercialism which reduces society to nothing more than a vast apparatus of production and exchange' (Durkheim, 1973, 44). Part of the problem in modern societies is that traditional sources of social regulation, such as religion, have failed to exercise moral constraint on an increasingly unregulated capitalist economy. More particularly, central to capitalist economies has been the emergence of modern commercial advertising, the aim of which is to create perpetual desire (Bauman, 1998, 23–41). It aims to produce wants that were not there before. When desires and aspirations are frustrated in modern societies, as they are most obviously during periods of rapid social change, individuals experience a loss of happiness and meaning (see Layard, 2006). Moreover, it is now apparent that this problem has been exacerbated by social media (Phillips, 1974, 340; see also, Hawton and Williams, 2002, 1374; Luxton et al., 2012; Pirkis et al., 2006; Pirkis and Blood, 2001; Yang et al., 2013).

These life experiences are frequently articulated in popular music, which serves to provide affective spaces within which listeners can reflect on their dissatisfaction and unhappiness. For example, Tori Amos's 'Maybe California' (*Abnormally Attracted to Sin*, 2009) invites the listener to eavesdrop on a conversation between two mothers, one of whom is contemplating suicide as a result of her perceived inability to meet her family's needs:

> ... please don't jump
> Why not? Nothing is making sense anymore to me
> I don't know when I stopped making him smile
> Now the kids see me cry all the time ...

The song, says Amos,

> is about the mother's role in our new world, where so many men are losing their jobs and the women are not, because they're cheaper. What is that doing to a home? If the mothers start breaking down then everything starts breaking down .... The idea that the mothers could really be contemplating jumping off a bridge or a cliff – that's coming out of a reality where a man is defined by being successful because he's a provider, and that's being taken away from so many homes. He's left there being stripped of this. And she can't give him that back. Wives and mothers live to make it better, and wives can't make this better. (Amos, quoted in *Polari Magazine*, 2009)

The same point is made by Alan Vega of the band Suicide in his powerful song 'Frankie Teardrop' (*Suicide*, 1977), which tells the tragic story of a poor factory-worker who 'can't make enough money ... can't buy enough food' and who is 'getting evicted' from his home. Eventually, he takes a gun and shoots his family before committing suicide. The visceral nature of the song is enhanced by dissonant electronica and Vega's distraught and anguished screams. The point is simply that Amos and Vega articulate some people's experience of a 'disturbance of equilibrium', which, as Durkheim (1952, 246) put it, 'is an impulse to voluntary death. Whenever serious readjustments take place in the social order ... [people] are more inclined to self-destruction'. Because Vega's and Amos's protagonists are unable to provide for their families, they experience a loss of meaning in life and, finally, choose death.

Richard Layard has also shown that habituation has an impact on levels of happiness. That is to say, those things we strive for, when realised, do not deliver the satisfaction we had anticipated. 'When I get a new home or a new car, I am excited at first. But then I get used to it, and my mood tends to revert to where it was before' 2006, 48). David Byrne articulates these feelings in the Talking Heads' song 'Once in a Lifetime' (*Remain in Light*, 1981):

> You may ask yourself, where is that large automobile?
> You may tell yourself, this is not my beautiful house
> You may tell yourself, this is not my beautiful wife ...

Throughout the song there is a sense of anxiety and dis-ease, as, one after another, goals that were once important, giving temporary meaning to life, fail to satisfy. As Layard comments,

> living standards are to some extent like alcohol or drugs. Once you have a certain new experience, you need to keep on having more of it if you want to sustain your happiness. You are in fact on a kind of treadmill, a 'hedonic' treadmill, where you have to keep running in order that your happiness stand still. (2006, 48)

Depression and Suicide in the Music Industry

A recent survey of 2,021 musicians (the majority of which were between the ages of 18 and 35) found that 68.5% had suffered from depression and 71.1% had suffered from high levels of anxiety (Gross and Musgrave, 2016; see also Preti et al., 2001). The report suggested that musicians are significantly more likely to suffer from poor mental health than the general public. Why? Another survey of 500 musicians found that lack of sleep, anti-social working hours, insecurity, constant travelling, poor diet, limited exercise, performance anxiety, fear of judgement, loneliness, and relationship problems all contributed to a deterioration in wellbeing (Help Musicians UK, 2014). As Alistair Smith of the band India Mill commented to me,

> I think a huge part of it that is overlooked are the massive highs and lows that come with creating, gigging and touring. I know that I was always in an awful mood after a great gig because you just want to be in that state permanently.

The problem, of course, is the 'no-one can sustain it so you then crave rest and silence'. This has an impact when it comes to 're-entry' into everyday life.

> You're just expected to get straight back into the swing of family life when you're head and body-clock has just been drilled into a certain pattern of life .... So aside from the dangers of the idealism that come with the creative side of music, I also think that the lifestyle

> of the brute logistics of the execution of it all contributes to frequent instances of depression and mental health issues. (Smith, 2019)

Perhaps unsurprisingly, therefore, popular music history it littered with stories of addiction, angst, and suicide, from Nick Drake in the 1970s to Ian Curtis and in the 1980s, and from Kurt Cobain in the 1990s to Scott Hutchison, Avicii (Tim Bergling), Keith Flint, Chester Bennington, and Chris Cornell in the last few years. As indicated above, often a contributory factor has been the 'rock star lifestyle'. As Alex Rosamilia has commented, prior to writing songs about angst and mortality, 'I didn't really have an outlet. My outlet was drinking' (Pearlman, 2019). Again, Syd Barrett of Pink Floyd famously used LSD as a

> magic ingredient that enabled him to dissolve the fortress walls of adulthood and skip gleefully back to infancy. From this vantage point he could invoke and reflect upon a world of certainty, a time before anxiety, before the expectations of ambition and the imposition of status .... (Chapman, 2010, 143)

As fellow band member David Gilmour commented, 'I just don't think he could deal with the vision of success and all the things that went with it' (quoted in Robb, 2006). While Barrett, of course, didn't commit suicide, he did suffer from angst and famously became reclusive after a short career.

More recently, James Arthur struggled with anxiety as a direct result of winning *The X Factor* in 2012. Having been made homeless at the age of 16, like Barrett, he found that, against his expectations, his musical success did not bring him the happiness he had imagined:

> I wasn't sure how to cope with the transition from being nobody, a person that didn't contribute to society, to someone who had a purpose. I had come from nothing, living rough or staying in awful bedsits. Then suddenly everyone had an opinion about me and I was staying in nice hotels ... I was so tired and I was trying to get to grips with my reality .... It's quite clear that money doesn't do anything for me, it's quite clear that fame doesn't do anything for me ... I just didn't know why I was feeling so unhappy ... I wanted to fall flat on my face and never wake up. (Arthur, quoted in Watts, 2013)

Again, following Durkheim, the experience of increasing detachment from others and the disappointment in relation to expectations or the inability to deal with the significant change in circumstances that success can bring can have a serious impact on an artist's mental health.

This angst is evident in the final words of Kurt Cobain, the lead singer of Nirvana, who, on 5 April 1994, put a shotgun to his head and pulled the trigger (see Cross, 2001). Near to his body lay a letter to his wife, Courtney Love, which articulated many of the characteristics of egoistic-anomic suicide. Describing himself disparagingly as 'an experienced simpleton', he outlines how he had become disengaged from society, from his family and from the music business. Consequently, he felt worthless:

> I haven't felt the excitement of listening to, as well as creating, music along with reading and writing for too many years now. I feel guilty beyond words about these things.

He continues,

> I can't fool you, any one of you. It simply isn't fair to you or me. The worst crime I can think of would be to rip people off by faking it and pretending as if I'm having 100% fun … I've tried everything within my power to appreciate it (and I do, God, believe me I do, but it's not enough) … I have a goddess of a wife who sweats ambition and empathy and a daughter who reminds me too much of what I used to be, full of love and joy, kissing every person she meets because everyone is good and will do her no harm. And that terrifies me to the point to where I can barely function. I can't stand the thought of Frances [his daughter] becoming the miserable, self-destructive, death rocker that I've become …. Please keep going Courtney, for Frances. For her life, which will be so much happier without me.[2]

More recently, on 10 May 2018, Scott Hutchison of the Scottish band Frightened Rabbit committed suicide after leaving a short note: 'I'm away now. Thanks.' His body was found in the Firth of Forth near to where he had written a song about suicide entitled 'Floating in the Forth' (*The Midnight Organ Fight*, 2008):

> And fully clothed, I float away …
> Down the Forth, into the sea
> I'll steer myself
> Through drunken waves
> These manic gulls
> Scream it's okay
> Take your life …

As these lyrics testify, a musician's angst, quite naturally, finds its way into the music. Alex Rosemilia expresses this process well:

> I decided I wasn't going to use alcohol, so it started coming out in songs instead. Man, that's fucking sad. I need to do something

---

[2]For a scanned image of the suicide note, see: http://bit.ly/2oKdpT5 (accessed on 08 August 2014).

> to deal with the stuff going on inside my head … I've always felt writing was very therapeutic. It's the ultimate form of catharsis. I've gone to therapy, like everyone else that feels that way, and that helps a bit. This kind of works in the same way. (Pearlman, 2019)

From here, of course, the worked through angst finds its way into the headphones of listeners. For many fans, this can be enormously helpful. However, when this shared angst is followed by the musician's suicide, the impact can be devastating. Hence, while the suicides of musicians are distressing, there has been some discussion as to whether they can lead to imitative behaviour.

The Werther Effect

'What is music if not a series of mothers?' It is, says the Norwegian musician Jenny Hval (2014, 71), 'about longing and belonging'. Her point is that music creates affective spaces within which the self can be understood, nurtured and accepted. The problem, of course, is that those listeners who have formed a significant bond with the music of a struggling artist – music that has functioned parentally and expressed their deepest longings and fears – are sometimes profoundly distressed when that musician commits suicide.

More generally, there is a widely documented negative effect of reporting celebrity suicides. As a recent study concluded,

> there is a need to remain vigilant about how suicide news is reported. Mental health professionals and suicide experts should collaborate with media professionals to try to balance 'public interest' against the risk of harm. (Pirkis et al., 2006, 2874; see also Yang et al., 2013)

Not only can suicidal ideation be increased by exposure to suicide-related material (Cha et al., 2010), but, in extreme cases, media reports can lead to copycat incidents of self-destruction (Stack, 2003). This phenomenon was first systematically analysed by the sociologist David Phillips, who referred to it as 'the Werther effect' after Goethe's novel *The Sorrows of Young Werther* (1774), in which the protagonist commits suicide. Although never conclusively demonstrated, it was widely reported that many devotees of the book imitated Werther's suicide. Hence, in some areas of Europe, the book was banned.

While Durkheim was unconvinced by the notion of suicide imitation, in 1974 Phillips cogently argued that, in fact, suicides do increase 'immediately after a suicide story has been publicized in the newspapers …. The more publicity devoted to a suicide story, the larger the rise in suicides thereafter' (Phillips, 1974, 340; see also Hawton and Williams, 2002, 1374; Hecht, 2018; Pirkis et al., 2006; Pirkis and Blood, 2001; Stack, 2003; Yang et al., 2013). Unsurprisingly, therefore, there is evidence to suggest that visits to hospital emergency departments 'for suicide attempts or self-injury increased following the announcements of celebrity suicides' (Jeong et al., 2012, 957; see also, Cheng et al., 2007, 69–75; Hecht, 2018;

Stack, 2003). For example, as Jennifer Hecht (2015, 158) comments, 'in the month following Marilyn Monroe's overdose, there was a 12 percent increase in suicides in America, with 197 more cases than usual'. It is now generally accepted that, in accordance with social learning theory (Bandura, 1971), 'the greater the amount of coverage of suicide in the media, the greater the increase in suicide rate' (Stack, 2003, 240).

Likewise, from an Asian perspective, King-wa Fu and Paul Yip make some interesting observations about the impact of the suicide in 2003 of the Hong Kong Cantopop star, Leslie Cheung. They found that 38% of their respondents had been adversely affected by Cheung's death. Indeed, there was a significant increase in the suicides of men using the same method as Cheung (Fu and Yip, 2007). Similarly, following the suicide of the South Korean K-pop singer Jong-hyun on 18 December 2017, and the attempted suicide of one of his Indonesian fans, the Department of Mental Health in Thailand released a statement warning parents to monitor their children's behaviour (Dam-Young, 2017).

This is interesting, because while there are, of course, cultural differences in attitudes to suicide, we should not ignore the transcultural continuities. Indeed, as the Italian sociologist Marzio Barbagli (2015, 307) discusses in his history of suicide, 'the differences between Europe and Asia have been diminishing for at least the past one hundred years'. In other words, while, as Foucault (1994) has taught us, we need to take account of the distinct *episteme* in which an event such as suicide is interpreted, nevertheless, there do seem to be significant psychosocial transcultural continuities concerning the relationship between a fan's angst and a musician's suicide.

Concluding Discussion

Focussing on Durkheim's analysis of anomic and egoistic suicides, the discussion has used both artist-centred and listener-centred approaches to explore a number of issues. Bearing in mind Durkheim's (1952, 279) point that 'self-absorption is not a good method of attaching oneself to others', arguably the most important questions to emerge relate to whether popular music can lead vulnerable listeners towards introspection, melancholia, and detachment. Although, the very act of suicide is individual and exceptional, and, as such, its causes are often difficult to ascertain, nevertheless, the issue of popular music's relationship to angst and suicidal ideation is an important one.

A 2013 longitudinal cohort study of the prevalence of deliberate self-harm and attempted suicide within the Goth subculture (which was fairly broadly defined) found that

> identification by youth aged 19 as belonging to the Goth subculture was the best predictor of self-harm and suicide attempt .... Self-harm could be a normative component of the Goth subculture including emulation of subcultural icons and peers who self-harm (modeling mechanisms). (Young et al., 2006, 1060; see also Silk, 2013)

This, of course, supports Durkheim's comment above. A larger longitudinal cohort study published in 2015 found that 15 year old teenagers who identified strongly with the goth subculture were three times more likely to be clinically depressed and were five times more likely to self-harm at age 18 than those who did not self-identify as goths (Bowes et al., 2015).

Having said that, and this is important, neither of these studies can be used to argue that being a goth causes depression or self-harm, but only that some young goths are more vulnerable to developing these conditions. Suicide and self-harm can also 'be explained by selection, with young people with a particular propensity to self-harm being attracted to the subculture' (Young et al., 2006, 1060). That is to say, the relationship could be due to social selection. Some young people with melancholic dispositions are attracted to music and subcultures that articulate how they feel. They come to the music already predisposed to self-aggression. A particular type of music attracts a particular type of person. Alex Rosemilia of Dead Swords makes precisely this point concerning his own musical preferences as a young person:

> Bands like Earth and Sunn O))) have been a big influence on me musically .... Both shoegaze music and heavier genres helped me cope with times when my mind was busier than I'd like it to be with thoughts no teenager should have. (Rosemilia, 2019)

Because of the way he felt, he was drawn to the music – not vice versa.

This last point is supported by a number of studies of heavy metal and suicide in the 1980s and 1990s. While some research found a link between the preference for heavy metal music and increased suicidal vulnerability (e.g. Graham et al., 1993; Stack, 1998; Phillipov, 2006), the reports frequently express caution about positing a direct causal link. For example, Karen Scheel and John Westefeld (1999) suggest that the reasons some adolescents listen to heavy metal and also self-harm can be traced back to personal characteristics and relationship issues, rather than there being any causal link between aggressive music and self-aggression. Likewise, Eric Lacourse, Michel Claes, and Martine Villeneuv have argued that the impact of a number of variables on the suicidal risk of heavy metal fans needs to be accounted for, including drug use, alienation, anomie, and difficult family relationships. Once these are taken into account, they suggest that 'to advocate for a causal relationship between musical preference and suicide would be misleading since no scientific data conclude in this direction' (Lacourse et al., 2001, 330–331).

Having said that, there may be a link between the music and certain behaviours, which, together, contribute to suicidal ideation. For example, Steven Stack and James Gundlach found a connection between country music and suicide, noting that, on the one hand, the lyrics 'portray drinking as a normative response to stress', but, on the other hand, fans 'are found to be at a significantly higher risk of suicide than nonfans in both in gun ownership and marital disruption' (Stack and Gundlach, 1995, 334–335). Hence, it is difficult to determine the extent to which popular music can be considered a contributory factor.

In the final analysis, the relationship between popular music, angst, and suicidal ideation is a complex one, in that while those with difficult lives and poor mental health maybe drawn to music that articulates their feelings, there is a question as to the overall impact of inhabiting 'affective spaces' (Partridge, 2014, 37–59) created by that music. So, while music might not generate suicidal ideation directly, it might nurture angst in vulnerable adolescents (Stack, 1998; Stack et al., 1994). For example, one study found that a person who has known someone who has taken their own life is more likely to be adversely affected by suicidal lyrics (Peterson et al., 2008).

Having said that, again, the above study also noted that some listeners responded to dark music with altruistic, pro-social thoughts of helping distressed friends. Again, another study, which highlighted the importance of gender differences, found that 'listening to music to express uncomfortable emotions could be a rather effective coping mechanism for girls since it reduces the risk for suicidal behaviors' (Lacourse et al., 2001, 331). Sad music can help some vulnerable listeners think through and articulate their feelings. Consequently, while it has been argued that there are 'well validated experimental manipulations that use sad music to induce symptoms of depression' (Bowes et al., 2015, 799), the evidence suggests that it can also have a calming effect (Yoon et al., 2019).

Moreover, as indicated above, some musicians, who have struggled with angst, specifically compose songs with vulnerable listeners in mind. A nice example of this is '1-800-273-8255' (*Everybody,* 2017) by the American rapper Logic. The title is the number of the National Suicide Prevention Lifeline in the United States. As some indication of its importance, the song has been described by the Lifeline as 'an unprecedented, watershed event,' which led, on the day of its release (28 April 2017), to 'the second highest call volume in its history.' Also, 'Google searches for the Lifeline phone number spiked significantly on April 28 by more than 100% .... Users searching for our phone number on Google often searched for it in conjunction with the terms "lyrics", "Logic", and "suicide" (Zisook, 2017). Interestingly, Alex Rosamilia has recently revealed his indebtedness to '1-800-273-8255', noting that his own song 'Ender' is inspired by it (Pearlman, 2019). Not only did it help him personally, but 'there's plenty of people who don't know how to fucking deal with [those feelings].'

In conclusion, while there is some evidence to suggest that popular music can contribute to cycles of self-harm in young people, individual responses depend very much on a number of personal factors, including the listener's relationships and mental health. Hence, while popular music culture (particularly the suicide of a musician) can become part of the problem, the positive role of music (both lyrically and musically) in enabling listeners to articulate deep feelings of angst and to explore of new ways of thinking and living also needs to be acknowledged and understood (Definis-Gojanović et al., 2009). Music matters!

Interactions and Sensoria

Joy Division. 1980. *Closer*, Factory Records.
Reed, L. 1992. *Magic and Loss*, Sire Records.

Suicide. 1977. *Suicide*, Red Star Records.

Wolfgang von Goethe, J. 1989. *The Sorrows of Young Werther*, Translated by M. Hulse, London, Penguin.

References

American Foundation for Suicide Prevention. 2018. Suicide Statistics. Available at: http://bit.ly/2PoQUyf

aNormaNonEmoKid. 2012. My chemical romance, *Urban Dictionary*, 27 August. Available at: http://bit.ly/2MS2HTV

Antonakakis, N. and Collins, A. 2014. The impact of fiscal austerity on suicide: on the empirics of a modern Greek tragedy, *Social Science and Medicine*, 112, 39–50. https://doi.org/10.1016/j.socscimed.2014.04.019

Bandura, A. 1971. *Social Learning Theory*, New York, NY, General Learning Press.

Barbagli, M. 2015. *Farewell to the World: A History of Suicide*, Translated by L. Byatt, Oxford, Polity Press.

Baudelot, C. and Establet, R. 2008. *Suicide: The Hidden Side of Modernity*, Cambridge, Polity Press.

Bauman, Z. 1998. *Work, Consumerism and the New Poor*, Buckingham, Open University Press.

Bowes, L., Carnegie, R., Pearson, R., Mars, B., Biddle, L. and Maughan, B. 2015. Risk of depression and self-harm in teenagers identifying with Goth subculture: a longitudinal cohort study, *The Lancet: Psychiatry*, 2(9), 793–800. https://doi.org/10.1016/S2215-0366(15)00164-9

Cha, C.B., Najmi, S., Park, J.M., Finn, C.T. and Nock, M.K. 2010. Attentional bias toward suicide-related stimuli predicts suicidal behavior, *Journal of Abnormal Psychology*, 119(3), 616–622. https://doi.org/10.1037/a0019710

Chapman, R. 2010. *Syd Barrett: A Very Irregular Head*, London, Faber and Faber.

Cheng, A.T.A., Hawton, K., Chen, T.H.H., Yen, A.M.F., Chang, J.-C., Chong, M.-Y, Liu, C.Y., Lee, Y., Teng, P.R. and Chen, L.-C. 2007. The influence of media reporting of a celebrity suicide on suicidal behavior in patients with a history of depressive disorder, *Journal of Affective Disorders*, 103(1–3), 69–75. https://doi.org/10.1016/j.jad.2007.01.021

Cross, C.R. 2001. *Heavier Than Heaven: The Biography of Kurt Cobain*, London, Hodder and Stoughton.

Curtis, D. 1995. *Touching from a Distance: Ian Curtis and Joy Division*, London, Faber and Faber.

Dam-Young, H. 2017. Indonesian fan attempts suicide following Jonghyun's death. *Asia One*, 21 December. Available at: http://bit.ly/2BQ2TwM

Definis-Gojanović, M., Gugić, D. and Sutlović, D. 2009. Suicide and Emo youth subculture: a case analysis, *Collegium Antropologicum: Journal of the Croatian Anthropological Society*, 33(2), 173–175.

Durkheim, É. 1952. *Suicide: A Study in Sociology* Translated by J.A. Spaulding and G. Simpson, London, Routledge and Kegan Paul.

Durkheim, É. 1973. Individualism and the intellectuals. In *Emile Durkheim: On Morality and Society*, Ed. R.N. Bellah, pp. 43–57, Chicago, IL, University of Chicago Press.

Foucault, M. 1994. *The Order of Things: An Archaeology of the Human Sciences*, Translated by A. Sheridan, New York, NY, Vintage Books.

Fricke, D. 1992. Lou Reed, Magic and Loss, *Rolling Stone*, 23 January. Available at: http://bit.ly/31Shand

Fu, K.-W. and Yip, P.S.F. 2007. Long-term impact of celebrity suicide on suicidal ideation: results from a population-based study, *Journal of Epidemiology and Community Health*, 61, 540–546. https://doi.org/10.1136/jech.2005.045005

Gaines, D. 1998. *Teenage Wasteland: Suburbia's Dead End Kids*, Chicago, IL, Chicago University Press.

Graham, M., Clarke, M. and Pearce, C. 1993. Adolescent suicide: music preference as an indicator of vulnerability, *Journal of the American Academy of Child and Adolescent Psychiatry*, 32(3), 530–535. https://psycnet.apa.org/doi/10.1097/00004583-199305000-00007

Greenwald, A. 2003. *Nothing feels good: Punk rock, teenagers, and emo*, New York, NY, St. Martin's Press.

Gross, S.A. and Musgrave, G. 2016. Can music make you sick? Music and depression. A study into the incidence of musicians' mental health. *Help Musicians UK, University of Westminster and MusicTank*. Available at: http://bit.ly/2qSRHNP

Hawton, K., and Williams, K. 2002. Influences of the media on suicide, *British Medical Journal*, 325, 1374. https://doi.org/10.1136/bmj.325.7377.1374

Hecht, J.M. 2015. *Stay: A History of Suicide and the Arguments against It*, New Haven, CT, Yale University Press.

Hecht, J.M. 2018. How the media covers celebrity suicides can have life-or-death consequences, *Vox*, 8 June. Available at: http://bit.ly/31U7V5O

Help Musicians UK. 2014. Professional musicians in the UK: health and wellbeing survey. Available at: http://bit.ly/2qJkp3k

Hval, J. 2014. The inner sleeve, *The Wire*, 366, 71.

Jeong, J., Shin, S.D., Kim, H., Hong, Y.C., Hwang, S.S. and Lee, E.J. 2012. The effects of celebrity suicide on copycat suicide attempt: a multi-center observational study, *Social Psychiatry and Psychiatric Epidemiology*, 47, 957–965. https://doi.org/10.1007/s00127-011-0403-7

Kernberg, O. 2009. The concept of the death drive: a clinical perspective, *The International Journal of Psychoanalysis*, 90, 1009–1023. https://doi.org/10.1111/j.1745-8315.2009.00187.x

Kposowa, A.J. 2003. Divorce and suicide risk, *Journal of Epidemiology and Community Health*, 57, 993. http://dx.doi.org/10.1136/jech.57.12.993

Lacourse, E., Claes, M. and Villeneuve, M. 2001. Heavy metal music and adolescent suicidal risk. *Journal of Youth and Adolescence*, 30(3), 321–332. https://doi.org/10.1023/A:1010492128537

Layard, R. 2006. *Happiness: Lessons from a New Science*, London, Penguin.

Levy, A. 2008. Girl, 13, Hangs herself after becoming obsessed with emo 'suicide cult' rock band, *Daily Mail*, 9 May. Available at: https://dailym.ai/34cFpy5

Luxton, D.D., June, J.D. and Fairall, J.M. 2012. Social media and suicide: a public health perspective, *American Journal of Public Health*, 102(2), 195–200. https://dx.doi.org/10.2105%2FAJPH.2011.300608

Malt, A. 2018. Approved: penelope trappes, *CMU*, 10 October. Available at: http://bit.ly/32W5pxe

Mental Health America. 2018. *The state of mental health in America*, Mental Health America. Available at: http://bit.ly/31Lzr5C

Moser, M. 2019. The issue with Billie Eilish's darkness. *Medium*, 22 April. Available at: http://bit.ly/348IbEi

NHS Digital. 2018. Mental health of children and young people in England, 2017. *NHS Digital*, 22 November. Available at: http://bit.ly/2MVyvYl

NME. 2008. My chemical romance speak about 'Emo' suicide, *NME*, 25 May. Available at: http://bit.ly/2BQWzoB

Osbourne, O. 2009. *I am Ozzy*, London, Sphere.

Ougrin, D., Tranah, T., Stahi, D., Moran, P. and Asarnow, J.S. 2015. Therapeutic interventions for suicide attempts and self-harm in adolescents: systematic review and meta-analysis, *Journal of the American Academy of Child and Adolescent Psychiatry*, 54(2), 97–107. https://doi.org/10.1016/j.jaac.2014.10.009

Partridge, C. 2014. *The Lyre of Orpheus: Popular Music, the Sacred and the Profane*, New York, NY, Oxford University Press.

Partridge, C. 2015. *Mortality and Music: Popular Music and the Awareness of Death*, London, Bloomsbury.

Pearlman, M. 2019. Why Alex Rosamilia has swapped gaslight anthem for doom and gloom. *Kerrang*, 6 March. Available at: http://bit.ly/2oseGyi

Peterson, R.J., Safer, M.A. and Jobes, D.A. 2008. The impact of suicidal rock music lyrics on youth: an investigation of individual differences, *Archives of Suicide Research*, 12(2), 161–169. https://doi.org/10.1080/13811110701857533

Phillipov, M. 2006. Self harm in Goth youth subculture: study merely reinforces popular stereotypes, *British Medical Journal*, 332, 1215–1216. https://doi.org/10.1136/bmj.332.7551.1215-b

Phillips, D.J. 1974. The influence of suggestion on suicide: substantive and theoretical implications of the Werther effect, *American Sociological Review*, 39(3), 340–354.

Pickering, W.S.F. and Walford, G. (Eds.). 2000. *Durkheim's Suicide: A Century of Research and Debate*, London, Routledge.

Pirkis, J.E. and Blood, R.W. 2001. Suicide and the media. Part I: reportage in nonfictional media, *Crisis: The Journal of Crisis Intervention and Suicide Prevention*, 22, 146–154. https://doi.org/10.1027//0227-5910.22.4.146

Pirkis, J.E., Burgess, P.M., Francis, C., Blood, R.W. and Jolley, D.J. 2006. The relationship between media reporting of suicide and actual suicide in Australia, *Social Science and Medicine*, 62(11), 2874–2886. https://doi.org/10.1016/j.socscimed.2005.11.033

Polari Magazine. 2009. From the fringes of the milky way: Tori Amos. *Polari Magazine*, 9 May. Available at: http://bit.ly/2NgNN8R

Preti, A., Biasi, F.D. and Miotto, P. 2001. Musical creativity and suicide, *Psychological Reports*, 89(3), 719–727. https://doi.org/10.2466%2Fpr0.2001.89.3.719

Robb, J. 2006. Syd Barrett, the Swinging 60. *The Independent*, 7 January. Available at: http://bit.ly/2Nm1t2r

Rosemilia, A. 2019. Interview with the author. 1 August.

Samaritans. 2018. Suicide statistics report. *Samaritans*. Available at: http://bit.ly/2parZns

Sanburn, J. 2013. Suicide in America: the people who answer the phone, *TIME*, 25 November. Available at: http://bit.ly/2NhOvCT

Scheel, K.R. and Westefeld, J.S. 1999. Heavy metal music and adolescent suicidality: an empirical investigation, *Adolescence*, 34(134), 253–273.

Sellgren, K. 2014. Young people 'feel they have nothing to live for'. *BBC News*, 2 January. Available at: https://bbc.in/2JHUsrR

Silk, J. 2013. Open a vein: Suicidal black metal and enlightenment, *Helvete: A Journal of Black Metal Theory*, 1, 5–20.

Smith, A. 2019. Interview with the author. 2 August.

Stack, S. 1998). Heavy metal, religiosity, and suicide acceptability, *Suicide and Life-Threatening Behavior*, 28(4), 388–394. https://doi.org/10.1111/j.1943-278X.1998.tb00974.x

Stack, S. 2000. Suicide: a fifteen year review of the sociological literature. Part II: modernization and social integration perspectives, *Suicide and Life-Threatening Behavior*, 30(2), 163–176. https://doi.org/10.1111/j.1943-278X.2000.tb01074.x

Stack, S. 2003. Media coverage as a risk factor in suicide, *Journal of Epidemiology & Community Health*, 57, 238–240. http://dx.doi.org/10.1136/jech.57.4.238

Stack, S. 2004. Emile Durkheim and altruistic suicide. *Archives of Suicide Research*, 8(1), 9–22. https://doi.org/10.1080/13811110490243741

Stack, S. and Gundlach, J.H. 1995. Country music and suicide – Individual, indirect and interaction effects: a reply to snipes and Maguire, *Social Forces*, 74(1), 331–335.

Stack, S., Gundlach, J. and Reeves, J.L. 1994. The heavy metal subculture and suicide, *Suicide and Life-Threatening Behavior*, 24(1), 15–23. https://doi.org/10.1111/j.1943-278X.1994.tb00659.x

Strong, C. and Lebrun, B. Eds 2015. *Death and the Rock Star*, Farnham, Ashgate.

Watts, H. 2013. James Arthur reveals his spiral into drugs and panic attacks after *X Factor* win. *Mirror*, 25 October. Available at: http://bit.ly/2Pq3pcE

Yang, A.C., Tsai, S.-J., Yang, C.-H., Shia, B.-C., Fuh, J.-L., Wang, S.-J., Peng, C.K. and Huang, N.E. 2013. Suicide and media reporting: a longitudinal and spatial analysis, *Social Psychiatry and Psychiatric Epidemiology*, 48(3), 427–435. https://doi.org/10.1007/s00127-012-0562-1

Yoon, S., Verona, E., Schlauch, R., Schneider, S. and Rottenberg, J. 2019. Why do depressed people prefer sad music? *Emotion*, 28 February. http://dx.doi.org/10.1037/emo0000573

Young, R., Sweeting, H. and West, P. 2006. Prevalence of deliberate self-harm and attempted suicide within contemporary Goth youth subculture: longitudinal cohort study, *British Medical Journal*, 332, 1058–1061. https://doi.org/10.1136/bmj.38790.495544.7C

Zisook, B. 2017. Logic's '1-800-273-8255' single has had a profound effect on suicide prevention awareness, *DJ Booth*, 13 July. Available at: http://bit.ly/2Jr9bqX

Discography

Tori Amos, *Abnormally Attracted to Sin*. Universal, 2009.
Dead Swords, *Enders*. Human Blood Records, 2019.
Frightened Rabbit, *The Midnight Organ Fight*. FatCat Records, 2008.
Logic, *Everybody*. Def Jam Recordings, 2017.
Metallica, *Ride the Lightening*. Megaforce, 1984.
Rites of Spring, *Rites of Spring*. Dischord Records, 1985.
Suicidal Tendencies, *Suicidal for Life*. Epic, 1994.
Talking Heads, *Remain in Light*. Sire, 1981.
Steve Vai, *Sex & Religion*. Relativity, 1993.
Amy Winehouse, *Back to Black*. Island Records, 2006.

Index

Note: Page numbers followed by "*n*" indicate footnotes.

Printed and bound by CPI Group (UK) Ltd, Croydon, CR0 4YY

13/09/2026

14961028-0002